THE MYTH
OF SCIENTIFIC
CERTAINTY

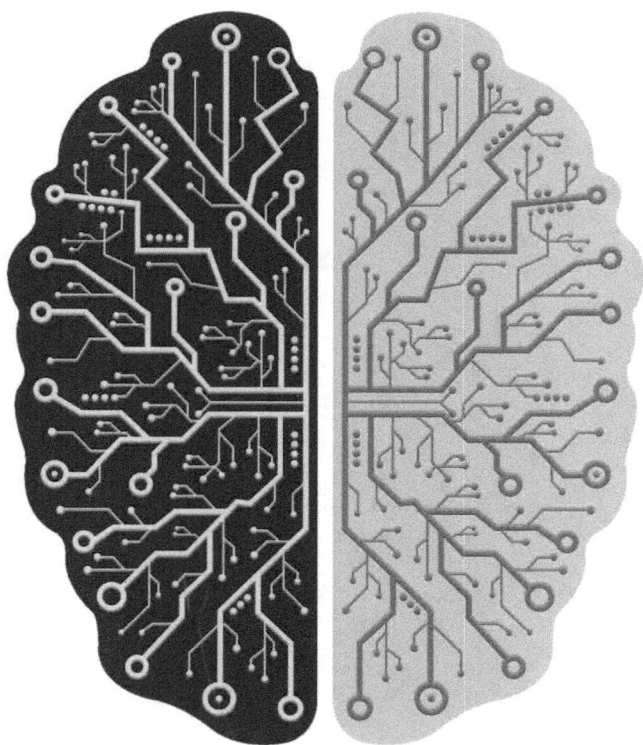

COLLIN BRENDEMUEHL

THE MYTH OF SCIENTIFIC CERTAINTY

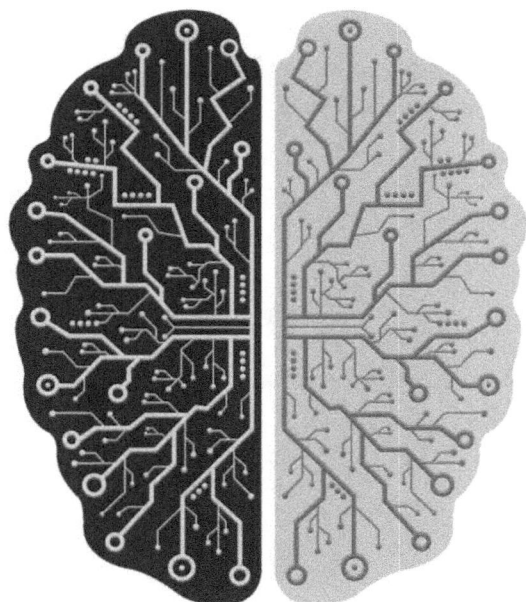

COLLIN BRENDEMUEHL

Published By

Paley, Whately Greenleaf Press

The Myth of Scientific Certainty:
  Scientific Theory and Christian Engagement
      by Collin Brendemuehl

Copyright 2018, Collin Brendemuehl, All Rights Reserved.

ISBN: 978-1-947844-39-1

Published by Paley, Whately, and Greenleaf Press,
    an imprint of the Athanatos Publishing Group.
          www.athanatos.org

Acknowledgments

For your support, encouragement, and assistance, a thank you to

Sarah Flashing
Chun-Liang Chen
Keith Plummer
Heather Zeiger
Dick Slemmons
David Koyzis
Maria Martin
Trevor Jennings
Doug Dunsmoor
Ryan Pflum

But most of all

Virginia

Table of Contents

Introduction

We live in the world of ideas. The world is not driven by economics or conflict or events as those are but the outcome of ideas. Science likewise is driven by ideas. There was the older idea that the world was composed of five basic elements (earth, air, fire, water, and ether). That was the science of the day and it was based on observation. The empirical approach – hands-on testing – is today's dominant approach. We study a world that we can see and measure and calculate. There are also some non-empirical methods, we call these models, that drive certain fields of study. These ideas are the product of a way of a set of more general principles that we call naturalism, empiricism, and scientism (among others). These principles see the world as a physical entity and generally dismiss theological concerns as unimportant. It is in light of these that the Christian who is engaged in the natural sciences develop a working knowledge of philosophy of science in order to find a place in this world of competing ideas.

The intent of this book is to provide some of the tools necessary to accomplish a singular goal: To separate naturalism from scientific practice so that the Christian might do scientific work without undue pressure. Naturalism is a set of assumptions that inform how many scientists approach their work. But one needs not become a functional atheist to participate in the natural sciences and examine the world around us. We have known for a long time that science can be conducted without a naturalistic worldview. We need not surrender to the pressure to adopt naturalism.

Accomplishing this goal will mean challenging certain aspects the current trends in scientific thinking. These trends may not represent scientific orthodoxy or consensus but, because of their popularity, rise to a level that lends that impression to readers. Then the additional questions that you are able to raise as a result of this effort will hopefully be ones which help produce better science and so dispel some of the myth of scientific certainty so common today.

This work is, at its heart, an introduction to and discussion of philosophy of science, accompanied by some of the current and popular principles attached to scientific inquiry. The reader who is not engaged in philosophy-type questions may find it tedious. I would ask you to keep in mind that this is intended to provide a stronger foundation for your scientific endeavors. It is here to help you be better at what you do because of that foundation.

The problem, as previously stated, is the attachment of naturalism to our methods of inquiry. In *Where the Conflict Really Lies*, Alvin Plantinga argued that naturalism and scientific inquiry exist in conflict. Dr. Plantinga worked to clarify the philosophical distinctions. Alternatively, the reality is that, in the workplace and in academia conflicts such as this frequently go unchecked. People go about their daily work without checking themselves for philosophical consistency.

One might easily dismiss these concerns as superfluous as long as the work is getting done. But that's the point – it is easy to forget to ask about how the work gets done and how the results are framed. We may be tempted to ask if our interpretation is somehow tainted or distorted. It is therefore necessary to provide a functional separation between the two. I hope to provide a solution for that concern.

Thus the purpose of this work is this: The Christian who wishes to work as a scientist, especially in the natural sciences, needs a bit of support. A current reason for this is the assumption by many that Christianity is incompatible with science. The popular mythos uses the term "anti-science" to describe Christians. Though we might readily dismiss it, the myth remains intact in the mind of the skeptic. A clear answer to the problem will assist the Christian in this arena. Thus the target audience for this project is the Christian scientist, the Christian student in the sciences, the Christian who teaches science, and the apologist who is engaged in a defense of the faith within the scientific community.

Some scientists see it as their mission to work on the front lines in an attack on theism. In the West the target is specifically Christian theism. Part of what drives this is the erroneous belief that Christian theism is somehow the enemy of science. Though from one perspective there is some element of truth to that (Christian opposition to naturalism has a definite effect on certain inquiry) it is not a necessary truth.[1] Yet the division remains. It is on account of this division that scientists will make statements such as these:

Jerry Coyne asserts that

> No creationist, whether of the Noah's Ark variety or otherwise, has offered a credible explanation for why different types of animals have similar forms in different places. All they can do is invoke the inscrutable whims of the creator.[2]

Donald Prothero similarly asserts:

> By now the evidence for evolution is overwhelming, so the burden of proof on the antievolutionists is much larger: they must show creationism is right by overwhelming evidence, not just simply point out a few inconsistencies or problems with evolutionary theory.[3]

The assertions of Dr. Coyne and Dr. Prothero challenge us to clarify the relationship between Christianity and science. Responses to this challenge have varied in both their scientific and theological considerations. But something else is taking place within this confrontation. It seems that many in society now view the Christian as being intellectually incapable of participating in science. Some go further and seek to isolate the Christian (or Christianity in general) from society. Those who reject or even question Darwinism and with it naturalism and materialism are at times isolated, marginalized, or removed from positions.[4] Of course this is not an every-day occurrence and I don't want to sensationalize the situation with fear mongering. It is, in fact, only a periodic and random situation. But it does happen and it reflects something about the changes in our culture.

The Questions We Want to Answer

We will work with these questions in mind: (1) Is the Christian able to work in the arena of science and still maintain an orthodox theology of creation? (2) What approach should the Christian take when facing this challenge? (3) What tools can the Christian bring to the discussion that will provide not only a defense of the faith but also a place for the Christian in a world that views us as ill-prepared for the natural sciences? (4) How can we counter any attempts at isolation and marginalization effectively?

When Christians are confronted with the challenge of what is called "science" (the challenge is actually naturalism, but more on that later) their responses usually fall into one of these three categories: *Ignore* the problem and stick with "faith," *disagree* with it by setting Scripture against it, or *capitulate* and give in to the arguments.

The first response is to completely avoid the question and the reply might be "Well, I'll just trust God on this." So the Christian goes on with "belief" or "faith" and will ignore the challenge.[5] What I think is really happening is that some Christians fear the question. It is a complicated and technical matter. But whatever the case may be the challenge is not confronted. It is perceived as a threat to the faith and thus is ignored.

We lose two opportunities with this approach. First, we lose an opportunity to improve our apologetic. If Christianity is everything we claim then perhaps there are answers. They may not be simple sound-bite answers for questions that pop up as more insults than clear thoughts, but answers they remain. Second, we lose an opportunity to engage the world. Again, if what Christianity provides has any substance to it then this is communicated interpersonally. Avoiding personal interaction diminishes our opportunities. This response ends up to our detriment and to the detriment of God's kingdom.

The second response, almost identical to the first, is to give a simple answer, perhaps a quote from Genesis 1:1. The person who does this tends to sound like the bumper sticker theology of "God Said It, I Believe It, and That Settles It." Though such faith may be commendable, the unwillingness to clarify to the hearer the foundation of one's faith (the substance behind the statement) is a clear weakness. In this instance the believer has gone from ignoring the problem to minimizing or trivializing it. Again, the issue goes unaddressed and the believer gains nothing useful to enrich his faith, to assist other believers, or to advance God's kingdom.

Of course not all who appeal to the Word are trivializing the threat. Many in the arena of Christian education, especially home schooling, and apologetics are confronting the issue. For example, the Creation Museum, while promoting young earth creation views also contains a good amount of scientific material[6] that is re-framed according to the young earth model. In this case, though, I found the treatment of the naturalistic models was as a threat rather than a challenge for learning and growing in one's apologetic and ministry[7]. (Such a defensive posture seems to me unproductive.)

For example, in the museum was (on my tour) a section displaying the negative social consequences of the relativism encouraged by scientific – naturalistic — ideals. Though I also accept this as true – that a godless naturalism leads to all sorts of social ills, it is also quite easy to fall into the "us versus them" trap, a belligerent and combative tone, and so always be at war with people. We might be tempted to treat the unbeliever as our enemy. This approach is not beneficial to the kingdom of God. If our ministry is one of reconciliation and redemption then a hostility-based approach seems counter-productive.

The third possible response is to acquiesce and propose that, perhaps, the natural approach was God's mechanism for creation. This is *theistic evolution* a view that has become a popular response to the challenge. But seldom (limited to my observations) do its adherents discuss the implications of evolutionary science to theology and ethics. Even more seldom do they discuss the point-by-point merits of evolutionary science or how the Christian might better pursue truth. In the end the doctrines of the faith are sacrificed to "reason" while many of its adherents sit unaware. Its impact on theology is undeniable and unavoidable, for if one accepts naturalistic evolution as the paradigm for the natural world then we must discard or radically modify all of historic doctrine to the point of incoherence. (How might one maintain, for instance, federal headship without a literal Adam?) Like so many matters in life, the choice that we make can be a two-edged sword and do damage in either direction if poor choices are made.

On the leading edge of this sword, the blanket acceptance of the naturalistic evolutionary model requires this effective rejection of some essential, historic Christian doctrines. Among these are original sin, the existence of God, Christ's resurrection, and the trustworthiness of Scripture. Without these Christianity becomes meaningless. But on the trailing edge, a blanket rejection of the particular evidence presented by the scientist might make the Christian and Christian theology appear irrational, unintelligent, and as such unsuitable for any application to the needs of the modern world.

This book includes a fourth response: To challenge the structure of some approaches to creating a model. To be more precise, the structure of the argument is not to be confused with the content of the argument behind the theory. The structure question addresses how the information is arranged rather than the information itself. This "structure" is the "scientific model" and is a way of framing information. But the content of the model is not the data. The content is the reasoning used to manage the information.

For instance, the fraud known as Piltdown Man is a piece of information. The same goes for the Brontosaurus situation[8]. Yet a challenge to these issues does not deal any serious blow to naturalistic science. The whole house does not fall if one brick is removed. The more effective approach is to examine the theory structures that have been built around the information and ask whether these structures are adequate and sufficient to answer the scientific questions. It is my proposition here that they are not. It is also the proposition of many scientists that this situation exists. Though my conclusions may differ the issue raised is the same.

The reasoning used for handling the information differs from the data. The use of analogy to move from a contemporary experiment to fill in historical data to complete a model is at times a stretch. This analogical approach has its problems. We will briefly delve into that matter as well.

The larger portion of this work is to examine the merits of the various scientific models. We will find come inconsistency and unsound science hiding amidst an equal amount of sound and valuable science. Again it is not a question of critiquing the information or how the information conflicts or how the idealizations (creation versus evolution or faith versus science) conflict. Rather the direction is to examine the efforts of a number of prominent and popular scientists and ask whether their models represent the data sufficiently and whether they are able to answer the objections of their opponents, especially those of other scientists.

After we have gained an understanding of science and examined the various models what we find is not the simplistic conclusion that Christianity and evolution are "compatible" in some obfuscated sense. To arrive at such a conclusion might put us on the road to compromising good doctrine and the principles of sound hermeneutics. Nor are we seeking to find some level of scientific compatibility between Christian teaching and science. The Bible is not a science book and treating it as such does it a disservice.

That is not to say that the Bible is not correct, but not all matters on which it speaks should be understood through our current approach to science. For instance, we know that the circumference of a circle is pi (3.14159...) times the diameter. But in I Kings 7:23 and II Chronicles 4:2 we read:

> He made the Sea of cast metal, circular in shape, measuring ten cubits from rim to rim and five cubits high. It took a line of thirty cubits to measure around it.

At first glance this looks like an error – and it would be if we were treating the Bible as a science or mathematics text. Pi is not 3, but 3.14. Mathematically this is an error of approximately 4.5%. But a common sense linguistic approach would give us a better solution. If the Bible were a mathematics book one would expect a certain level of precision. But the Bible is not a textbook; it is, in large part, a history book.[9] Part of the content is a record of conversations. Today, we often speak of 14 minutes after the hour as "quarter after," and of 16 minutes before the hour as "quarter 'til," even though neither are exactly a quarter-hour from the nearest hour. Like this, the Biblical example is frequently a record of common and casual language. To say anything more regarding language is unnecessary. (Thus, to complete a proper interpretation, the value 3 was rounded for ease.) In this example a proper plain language treatment of the Bible's content gives us the proper conclusion. With respect to the matter of scientific questions we do well to avoid reading into Scripture more than is there. When we do this our expectations become unrealistic. Should our perspectives about what we then come to think that the Scriptures ought to give us for answers are not met then one's faith might be affected negatively, for all the wrong reasons.

Just as we should look at the Bible through a proper lens and see it on its own terms, likewise we will examine science and naturalism, each on its own terms. The first section of this book will concern itself with the fundamentals of theory. We begin with a definition of what constitutes "science." Though this may seem unnecessary, how the term is defined has an impact on what is found as acceptable and legitimate. For instance, does a theory such as tachyon theory, which has yielded no verifiable results, constitute science? For some the answer is yes and for others, no. This divergence on whether a theory without fruit constitutes science has some effect on scientific theory and even on education. It affects how science is taught and what students are instructed to assess as legitimate and acceptable knowledge.

The definition of science is followed by an examination of the various definitions of evolutionary theory[10]. It is this branch of the natural sciences that causes Christians the most problems and as such is the one that requires analysis.

The term "evolution" has its nuances, first from any notable change or adaptation and speciation, then going further into sociology, as well as cosmology. Not all scientists accept the breadth of these definitions of evolution and are, I think, rightly critical of the abuse of language and over-application of the theory.

The definition of evolution is also nuanced by way of the models employed to describe the evolutionary process. There have been a number of models proposed over the past two centuries. Some of these have been generally rejected, such as directed panspermia, which posits that life on earth was intentionally planted here by other beings from other planets. The more conventional theories remain, including adaptation, random genetic change, a synthesis of adaptation and genetics, and more recently through a genetics-driven, information-driven (even mathematical) process. These structures account for currently accepted scientific and evolutionary orthodoxy, though one might properly consider each segment a type of "denomination" within the scientific community.

Section two provides an overview of the various evolutionary models will be presented, allowing the reader to gain perspective on the character of evolutionary theory. It will become evident that, although scientists might agree on what may have happened there is no consensus regarding how it all actually happened. There is no agreement on whether behavior develops fitness for survival (and with that fitness the genes take on new features), whether a "random" genetic change adds to fitness thus advancing the species, or whether these events came about through some combination of the two.[11]

Section three deals with how a number of scientists argue. Their arguments contain certain assumptions and among these is a thing called directionality. Directionality is the principle that the engine of biological development must always lead to the advancement of a species. In addition to directionality the scientist might appeal to vague generalizations where the model in use has failed to answer a challenge. One of these is theories that have been built around the development of visual capacities. The answers you read from some scientists on this subject may leave you wondering whether there is a consensus on the issue or if "just because" has become a new dogma.

Section four will provide the Christian with ways to challenge the naturalistic approach to science. This is the section in which Christian apologetics takes the offensive. I present no model to replace the naturalistic model. It is not my goal is not to undercut these models *per se*, but to undercut their attachment to naturalism. While sections two and three may seem to be disconnected from this because they deal with theory details, they also deal with the assumptions that go into scientific structures. These assumptions are the concern.

The Christian, with sound answers to the challenges presented, may then participate in the natural sciences in good conscience, working alongside colleagues who hold to naturalism (in either research or academia, even though co-opted by naturalism) and at the same time reject naturalism. Thus the particulars of how a cell or gene behaves need not be a stumbling block for the Christian as this is a matter separate from a naturalistic worldview. For one to acknowledge biological development and evolution within the created order would do no harm to one's theology. That is, genetic behavior is what it is no matter whom is examining the data. The believer can work without intellectual conflict or a conflict of conscience by placing this behavior under the umbrella of creation instead of under the umbrella of naturalism.

Information is always processed through the filters of experience, history, and training (and of course our fallenness), and as such is always tainted. Culture provides a shared coloring of information for those who live in that cultural context. For instance, in every nation the concept and enforcement of justice is viewed according to the laws of that nation and culture. Is it just to cut off the hand of a thief or to execute the adulteress or the rape victim? The answer would be yes in many nations, but in the United States and many others it would be no. [12] Different cultures and worldviews bring views of truth which differ drastically from our Western, Rationalist approach.

In this light I find an information-based apologetic regarding naturalism and the question of origins to be less than sufficient. If we win arguments regarding information details but do not deal with this worldview issue then the worldview will seek other supporting mechanisms. (Naturalism will simply adapt by acquiring a new or modified model but itself remain intact.) As a result we should have greater success in this area by dealing with the meta-questions of worldview.

This apologetic is also not intended to defeat naturalism. (Alvin Plantinga has done a fine service in that arena in his work *Where the Conflict Really Lies*.) The goal is instead to keep the door open so that the false sense of certainty purported by many who espouse naturalism can be dealt with by undercutting the equally false sense of certainty about the mechanism (materialistic evolution) which supports it. Materialistic evolution serves as the apologetic for naturalism but materialism as a guiding principle is insufficient to explain today's world. Christianity provides the best non-materialist explanation.

The concern here is not the details or the scientific particulars, but the broader framework, the scientific approach to framing the evidence to fit the model. This is not an attempt to dismiss the good scientific material that is produced by scientists. This is also not just a review of the presuppositions that drive naturalistic science. This is about the model, the working assumptions that the scientist uses to support the premise of naturalism. It is through some of naturalistic evolution's weak model structures that we can effectively challenge and so leave room for confidence for the Christian working in the natural sciences. [13]

I. Toward A Functional Definition of Science

What we are defining first is science as the study of the natural world, i.e., *natural science*. Once the term is understood in its fuller sense then one will be better able to understand the character of scientific inquiry in general.

Some Guiding Principles

Defining "science," it appears, is like trying to define "love." The term has taken on many different meanings and a discussion about which one is in use at any given time seems prudent. One hundred years ago the tendency might have been toward defining science as *empirically testing the physical world* but today a myriad of other definitions take their respective places. If science were simply the requirement for methodological study, then the scope of science might include all of research – including questionable areas such as UFOs. Alternatively if science were only a test for measurable results then it would be necessary to remove many theoretical inquiries from the scope of the definition. If science is reduced to problem solving, then there is yet another nuance that removes any seemingly unproductive work from the definition. Finally, if science is understood more broadly, if we can harmonize several suitable definitions, then a comprehensive definition can be built.

A Case for Pessimism?

There also is the question as to whether or not scientific inquiry is even viable. How we understand the world around us, as well as understanding why and how people understand anything at all, gives us cause to inquire about method and approach to science. It appears that all knowledge may be tainted by culture, assumptions, and presuppositions and as such comes into question[14]. Think of it this way: Take the clause "Based on what we know now …" into any generation. For that generation that clause serves as a sort of qualifier for the legitimacy of its scientific inquiries. Alternatively, when we look at scientific inquiry in the past we proceed to modify the qualifier accordingly: "Based on what they knew then …" It seems obvious to us in our modern world that we know more today and have a better system of learning and analysis. The result is that we, sometimes arrogantly, denounce past work as unscientific.

The pessimism of this qualification gets clearer: If we know, and assume to know with a high degree of certainty, that what was done in the past was not correct, what is there today to make us think that we have any right to be smug? Are we justifying error based on a lack of evidence, tools, or technique? Might tomorrow's more-precise inquiry render today's meaningless? What is orthodoxy today may be tomorrow's heresy.

Inductive Reasoning in Science

Inductive reasoning is an estimate of the likelihood of something being true or correct. Because induction leads to likely conclusions it thus does not lead to fixed absolute conclusions. Consider this: Not all, and probably very little, scientific study will produce certain or incorrigible conclusions.[15] That is, the tests can always be improved, the constraints tightened, and thus the results that are reached will become more precise.

What is today accepted as fact may very shortly be shown as an error. As a result the conclusions of inductive inquiry are not to be held onto as though they are permanently established. Scientists accept that. It's just the nature of experimentation and information evaluation. But here we are more concerned about the character or structure of a theory than its fruit.

Limiting the Scope of Inquiry

Physicalism (at its most basic level) is the first limitation. It means that we are only studying the physical world. Note here that physicalism is not the same as *naturalism* and does not carry the same demands.[16] Naturalism makes metaphysical statements and physicalism seeks to avoid those. For instance, physicalism does not mention in any way the broader questions of *why* though it may address *how*.

On the other hand, something called *methodological naturalism* insists that there is nothing metaphysical that is testable. It introduces metaphysical questions (*why* questions) into the conversation. Though this approach seems to functionally differ none from physicalism it should not be missed that it also gives some weight to those why questions.

Then there is *metaphysical naturalism* which demands that there are no *why* questions. This approach makes a clear demand, that there is nothing metaphysical and only the natural world exists. It thus does not allow the why questions, especially not theological questions.

Frameworks of Scientific Theory

Our starting point for scientific inquiry begins with physicalism, which is generally empirical in its approach. This practice is encompassed in the "scientific method" which has certain rules to be followed. These rules are found in something called the Received View. The Received View is the foundation for empirical inquiry. After that come the modeling and mechanism/schema approaches to scientific inquiry. Models move us from a concrete set of rules to something more abstract, something less empirical.

On the Received View

The Received View is a set of rules that apply to empirical inquiry. A proper theory depends upon rules and these rules are to be strictly adhered to. Following these simple rules helps create a proper (fully-formed) theory and compliance to these rules determines whether or not a theory is properly scientific. Though a bit complicated in appearance, the (abbreviated) rules are as follows:

(1) There is a first-order language L (possibly augmented by modal operators) in terms of which the theory is formulated and the logical calculus K defined in terms of L.

(2) The nonlogical or descriptive primitive constants (that is, the "terms") of L are bifurcated into two disjoint classes …

    Vo must contain at least one individual constant

(3) The language L is divided into the following subcalculi:

    (a) The observational sublanguage

    (b) The logically extended observational language.

    (c) The theoretical languages.

(4) Lo and its associated calculi are given a semantic interpretation which meet the following conditions

    (a)The domain of interpretation consists of concrete observable events, things, or things-moments; the relations and properties of the interpretation must be directly observable.

    (b)Every value of any variable in L_O must be designated by an expression in L_O.

(5) A partial interpretation of the theoretical terms and of the sentences of L containing them is provided by the following two kinds of postulates: the theoretical postulates T (that is, the axioms of the theory) in which only terms of V_T occur, and the correspondence rules or postulates C which are mixed sentences. [17]

To put it more simply, the Received View defines the *structure and rules* of a theory. It defines the construction of the formula and the relationship of its content to the subject matter. It is meant to certify that the theory is clear and stated with sufficient precision to allow processing and testing, that the content of the theory will be consistent with the content of the anticipated and measurable conclusions, and is stated in a precise enough manner so as to allow the results to be verified. The terms control the language that comprises the theory and it controls the statement relationships within the theory. It does not define the outcome of the theory (whether it is successful at reaching its conclusion) and it does not determine whether fact or error is produced by the theory.

The Received View is not the scientific method that we learned in school. It is a set of rules that determines how a test is set up by way of the scientific method. It says that a test for gray squirrels should not result in a conclusion that describes squirrel*ness*, red squirrels, or gophers. The language of the test, the hypothesis, and the interpretive method should all be clear, specific, limited in scope, and inter-related.

Let's apply it to a theory that uses the scientific method. It might go something like this:

1. Observation/Problem	Are all triangles 3-sided?
2. Hypothesis/Theory	Is this object 3-sided?
3. Test/Experiment	Count all sides of the object
4. Measurement	Add up the numbers
5. Evaluation	Do the numbers total 3?

Yes, report results as successful

No, report results, revise theory and retest

The Received View says that the language throughout the process must to be consistent. If one *postulates* three sides then one *theorizes* regarding three sides, one *tests* three sides, one *measures* the number of sides, and one *reports* regarding the number of sides. This constraint limits the language and the tests to a clear correspondence. The constraints placed the test, as gathered from the Received View, are here useful and functional to arrive at the test results. Thus, to be scientific is to have a *correct structure*, not a certain type of product or outcome or even a successful execution. But if one were to modify the postulate (and subsequent theory and test) to look for three sides on a square, and though the results would prove false, it might still be a valid theory structure.

The structure of our test was designed and followed consistently to lead to a conclusion. It is not required that a test lead to a necessarily true conclusion. It is equally legitimate to test for a false or erroneous conclusion (will 'x' always fail under condition 'y'?) if that is the goal established up front. Any test that leads to inaccuracy is still useful for other purposes, such as clarifying what is not (showing what cannot work), as opposed to what is (showing what does work).

How would a theory test for what is not or what does not work? A return to the triangle test and modify the parameters will add some clarification:

1. Observation/Problem	Triangles are not 4-sided.
2. Hypothesis/Theory	Is this triangle 4-sided?
3. Test/Experiment	Count the sides
4. Measurement	Add up the numbers
5. Evaluation	Do the numbers equal something other than 4?

Yes, report results as successful

No, report results, revise theory and retest

At this point we're entering the inductive nature of inquiry and departing from what is considered deductive reasoning. We are looking at the probability that there might somehow be a 4-sided triangle. Though we know *a priori* that there can never be such a thing, the theory structure is what makes that allowance. Let's take it a step further and make it even clearer:

1. Observation/Problem	Does this square have 3-sides?
2. Hypothesis/Theory	This square has 3-sides.
3. Test/Experiment	Count the sides
4. Measurement	Add up the numbers
5. Evaluation	Do we reach the value 3?

Yes, report results as successful

No, report results, revise postulate and retest

The theory might be considered sound in one sense because one will always be able to reach the value of three when counting the sides. Because we will always reach three it might be concluded from the evidence that squares have three sides. It's true. They have three, but not *only* three. The obscurity allowed by not establishing sufficiently narrow or adequately precise parameters (language) can leave the definition of "square" open for debate. That's one of the problems that arises from an inadequate set of constraints.

All three of these are legitimate theory structures but have a qualitative difference because of the language issue. The precision of the language is everything. Language that does not deal with missing pieces of evidence can create a false-positive result.

The product of a scientific endeavor thus conforms to the language (as defined by the criteria of the Received View) of that science. This is true when one depends upon this approach to define the character of that inquiry. It holds a certain level of legitimacy. It is only expected that theories that find their home within the language of naturalism would produce a fruit suited to the language of naturalism and theories that find their home in the language of theology would produce a fruit consistent with the language of theology.

These rules were created principally for evaluating the physical and observable world. They were also designed to exclude metaphysical discussions.[18] Their application coincides with a physicalist approach to scientific inquiry.

One notable weakness is that the Received View has no clear falsification category. It has no position at which the outcome can be either disproved or qualified in terms of error or shortcoming. Instead it depends upon the accuracy and precision of the Language and Criteria to make false results discernable. This feature came by way of Karl Popper[19]. Though here we will leave questions about Popper's approach aside, except to say that falsification introduces a metaphysic that leaves some uneasy. A more developed version of this would include the necessary negative calculi, which would force Language precision and so create the appropriate falsification language and tests. For instance, if formula requires both observations A and B to be true then the V_O (the statement about our test) would list $(A \cdot B)$ (that is, both A and B must be true). But to be certain that an error is dealt with correctly, a matching and contrary criteria would be required.

The formulation is standard but there is a semantic and contextual issue for the formula. If we are testing for failure then we use $(\neg A \lor \neg B)$ (if either A or B is false then we have falsified the result) but if we are testing for success in passing the negative test then we use $\neg(\neg A \lor \neg B)$ (if neither A nor B is true then we have falsified the result). Either way the product is the same; only the Language will change according to the semantic. The inclusion of this negative check forces the criteria (A and B) to be stated clearly enough that they can also be stated clearly in negative terms. "If X is observed then A succeeded," is contrasted with "If X is not observed then A fails." These constraints also hold true for V_T (the terms used in the initial theory and consistent with that language).

The Received View shows some other shortcomings, as it is inapplicable in certain situations and for certain other sciences. Suppe acknowledges this:

Thus, without attempting to delimit the class of theories that properly qualify as being scientific, *I will establish that not all scientific theories admit of the canonical axiomatic formulation required by the Received View.* Then I shall argue that some theories do not admit of the required canonical formulation, thereby establishing that the Received View is plausible for some but not all scientific theories.

To demonstrate that not all theories commonly referred to as scientific admit of the canonical reformulation required by the Received View, it will suffice to show that some of these theories cannot be axiomatized fruitfully. The question here is not whether all theories can be axiomatized – for it is the case that any theory, scientific or otherwise, can be "axiomatized" in a trivial fashion by the mere listing of the symbolizations of all known results – but rather whether they all can be axiomatized fruitfully.[20] (italics mine)

What he said simply is that a scientific theory may be proper and fruitful without following the Received View completely. There is some room for changes in language to add breadth and depth to a theory. I tried to give a little hint of that in the squares with three sides test. The constraints might have been tightened by saying "only 3" or "4" or something else. Likewise the nature of triangles changes if we employ a Z-axis so that we might test for a triangle with three 90-degree angles. The allowance of broader language adds more than just constraints. It adds context to a test. We are free to add more to a test than just the data and constraints. This opens the door to other constructions of scientific theory that remove from inquiry that naive sense of brute fact objectivity, making inquiry about more than measurable results.

What we can see, then, is that not all science is as objective as it might appear. Objectivity requires a remote position for external observation and the introduction of language might set goals beyond what seems to be reflected in the data. The Received View, especially when held to strictly, works very well for the observable and measurable but what of non-empirical matters such as theoretical physics? How do we verify results from either broader language or non-empirical tests?

How do we know that a theory works? How are results verified? The Received View might be considered the embodiment of *verificationism*. This is the constraint that the meaning of an empirical sentence was given by the procedures that one would use to show whether the sentence was true or false. If there were no such procedures then the sentence was said to be empirically meaningless.[21]

So while the empirical system has verification (through measurement and reporting) as part of its structure, and with it a certain level of objectivity, other approaches may not have that benefit. The door is again open to discussing matters of apparent objectivity that were raised earlier. Objectivity is a position to which we might aspire but it has been questioned by many whether or not we can attain any objective position at all. For if there is no clear sense of understanding then how might we ever find any objective conclusions?

And This Means

I've taken a back-door approach to address the requirement for theory structures that goes beyond the observable and measurable and into the field of the non-empirical. There are certain areas in science that begin with theory structures that cannot be proven with the then-extant technologies and may not be proved for many years, if ever. For example, during the last century the fusion needed for the hydrogen bomb was theorized many years before the bomb was created. In this theory was the postulation of atomic elements and how they would interact under the proper conditions. The theory worked. So the question follows: Was the theory not scientific from the beginning or did it become scientific once it bore fruit?

A current example of this type is tachyon theory[22], derived from the apparent logical necessity of a thing's existence without ever completing measurable experiments with fruitful results. This theory remains on paper only yet by many is accepted as valid scientific inquiry.

Whether one approaches theories by way of the Received View or if one discerns further to treat probabilistic or deterministic theory against corrigible or incorrigible data, there is still the question of objectivity and how both data and results are perceived. Patrick Suppe's alternative argument depends upon a perceptual certainty[23] and that creates a serious concern, for at this point we are forced to leave the theory structure (and the assumption of some level of objectivity) and delve into the matter of warrant and justification. The implication, it appears, is that a theory is not so freestanding and objective, not so separate from the subjective as we might like to believe.

An initial response to this matter is that it is impossible to separate the natural science from "religion"[24]. Despite the efforts of Hume and Kant[25], the two are inextricably tied. As Seager says:

> The positivist theory of meaning has fallen (far) out of favor nowadays, but the distinction between the two contexts of scientific activity remains an important bulwark against metaphysics. Yet I believe that even a cursory examination of modern science will show that there is no easy way to separate metaphysical from "purely scientific" doctrine unless one is willing to swallow a radical (neo-empiricist) vision of science which itself embodies a metaphysical position. Furthermore, if we follow this route, the legitimate place of science within our culture is sadly diminished. [26]

Sociological implications aside, it is clear that the natural sciences do not escape the question of metaphysics, making it necessarily religious, which is consistent with Clouser's conclusion [27], and so comes under the constraints of metaphysics. Science, it appears, is frequently as concerned about metaphysical questions as it is about natural ones.

Despite these criticisms positivism (the idea that only the verifiable is acceptable as legitimate knowledge, and that means the exclusion of all metaphysical questions) remains alive and well today. In this fashion Gould looks at scientific theory by way of an outcome that must happen according to some law – that somehow all of this was, according to the theory, a measurable, predictable, and expected outcome.

> However, when we move to the species level, the analogous driving phenomenon of *directional speciation* suffers no constraint or suppression – and may represent one of the most common modes of macroevolution. Two major reasons underlie the high potential frequency for directional speciation … First, as noted in several other contexts, the species-individual does not maintain integrity (as the organism does so by suppressing differential proliferation of some parts over others. … Second, since new species-individuals must arise with sufficient heritable novelty to win reproductive isolation from their parent … all species births include genetic change as an automatic consequence. Any statistical directionality in such changes among species in a clade will produce a trend by drive.[28]

This is a problem for Gould. Even though he sought a naturalistic solution he ended up with a metaphysical issue – teleology. His theory developed into a being with a purpose. Of course that purpose was not one of mental intent. But it was one where the biology worked toward its own ends. The assumption that this trend could happen, that there is a trend toward developing better/stronger/faster entities, is not a scientific question. It is a metaphysical question.

The reply to all this is not to say that both naturalistic theory and intelligent design, because of their metaphysical assumptions must and must *automatically* insert the theological (i.e., the metaphysical) into the discussion and, in opposition to this, that "real" science does no such thing. The complaint is, in fact, quite the opposite. Current scientific theory, with its foundation in *naturalism* and evolutionary theory with its teleological metaphysic, has the propensity to insert its own "religious" concepts (its own theology of a sort, though it is clearly not "theism") into the discussion. And herein is the conflict: Though the stated premise of naturalism is the removal of anything theological from all discussions, it also makes a clearly theological statement as part of the theory-making process.

What makes metaphysics a religious, even theological, question? Both share the assumption that there is something outside of the physical world. Gould called it a *trend*. Some thought, some first principle, some idea is taken as axiomatic. That might be dialectical materialism, special creation, or even uniformitarianism, depending upon one's worldview. Each of those is held by someone as axiomatic. Thus they fit a broad, non-ecclesiastical definition of religion, matters on which all else depend.

But the scientist (or student) wants to practice physicalism. The goal is to get the work done and not worry about these religious and metaphysical questions. It would be easy to become conflicted in one's mind and dismiss the question in favor of getting the work done. Unfortunately that does not make the problem go away. I would suggest that inquiry improves by struggling with the question rather than avoiding it. It is then that one sees the metaphysical assumptions written into the theories.

Now we see one of the problems with the assumption of objectivity. Even a return to just working from a physicalist approach is not enough. Physicalism is not a simple alternative to (methodological) naturalism. This may work for some of the physical sciences but is inadequate for the more abstract scientific ventures and misses the metaphysical questions that bear some weight on a theory. Science seems to require some metaphysical assumptions.

Model Theories

What we think of as "good" science may produce false, inadequate, or obscure results. Take germ theory for example. A century ago we came to understand the basics of germ theory and the conclusion that "germs cause disease" was enough for the times. But the science changed over time and the term "germs" has been expanded to include bacterial, viral, and other elements. "Germ" was a useful and even practical explanation at the time but today it would be laughable to employ such a broad and obscure term within a scientific endeavor. Good science came to a conclusion that was precise enough for the time, as precise a conclusion as could be understood at the time, but which is totally inadequate today.

The simple empirical approach was not good enough. For this reason and more the *Received View* is now the *Once Received View* [29] (or ORV) and other frameworks now take their place in theory making. It is no longer the standard for scientific orthodoxy (if there ever was one), if there even is any orthodoxy within scientific methodologies.

While the Received View may have a place in empirical testing of the physical world there are some tasks it is incapable of addressing. It is a system concerned with particulars and not with broader operations and processes. It is not a modeling system whereby a process or system can be duplicated or emulated. And it is not capable of providing a complex schema. These next two systems provide solutions in those areas. To help explain these systems, each is presented with an every-day example or parallel of how the theory structure might be employed outside of the world of science.

The Model Approach (aka The Model Model)

One way to deal with the precision required of the Received View was to insert abstraction in place of the Received View's precision. Sort of. Instead of building a list of successive tests the scientist builds a model and processes to run (to test) against this model. [30] Models are abstract theories rather than precise measurements. The theory moves from being a linguistic structure to a semantic one. This approach has the advantage of removing the language constraints, thus forcing them to the desired level of abstraction. It allows a scientific theory to test against a principle instead of a precise outcome.

A variety theories are built on this method. All historical theories use semantic inferences that lead to the best explanation. Weather forecasts build on data both past and current to make statements about what may likely come about.

Here we cease to use the rigid linguistics that wrap the scientific method and we change to something more semantic, the language of models. The process of refining the semantics is one of the processes used to improve a model. Like the Received View's potential problems with unclear theory, the necessity to control the obscurity of the abstractions in the Model[31] presents the same issue.

One weakness is that the abstractions can become too abstract, too removed from producing a clear conclusion. A model that is too broadly stated is meaningless. Craver clarifies the need for some sort of fruitful outcome:

> Each of these MM [Model Model] approaches to laws provides tools to grapple with issues of universality, scope, abstraction, and idealization. Suppes' approach is *prima facia* more appealing because it sustains the reasonable claim that theories express empirical commitments. Neither approach clarifies the necessity of laws. Giere (1999, p. 96) suggests that the necessity of laws statements should, like issues of scope, be considered external to theories. This suggestion is unattractive primarily because many uses of theories (including explanation, control, and experimental design) depend crucially upon notions of necessity; an account of theories cannot cavalierly dismiss problems with laws precisely because laws (or something else filling their role) are so crucial to the functions of theories in science. [brackets mine] [32]

Abstraction is thus not to be separated from the rules. Just the opposite is true, in fact. It's just that there are different rules for models than for the Received View.

This approach to abstraction – coupling rules with the process – has found a parallel in software development. Software construction is often done in terms of an abstract routine that processes data by way of a predicted format, accompanied by the appropriate exception handling, using generalized routines as (opposed to the traditional and seemingly empirical top-down approach where information is simply taken in steps, piece by piece). Using a very simple example, the resulting algorithm for a routine has changed from

Routine
      If A is true then do X else do X'
      If B is true then do Y else do Y'
      If C is true then do Z else do Z'
End Routine

to the slightly more abstract

Class
      ProcessA(A) { }
      ProcessB(B) { }
      ProcessC(C) { }
End Class

For those not familiar with programming languages what has happened here is that a step has taken past the straightforward "if-then-else" approach. Units "A," "B," and "C" each contain more than just a "true" or "false" value. They have become abstract entities that may contain multiple data elements and even code. This allows for "data" to initiate and respond as well as be the placeholder of information. It is complicated to be certain.

This new type of abstracted structure is implicit in object-oriented software design where the old data elements that might have represented keystrokes or numbers now represent both data and processes. This is especially true in Java and C++[33].

Unlike a software development model, other models may be descriptive rather than prescriptive. A familiar scientific model of this character has to do with disease and inoculation statistical studies.[34] A disease, like influenza, may spread rapidly if few or none of the populace is inoculated against it. But there comes a percentage point where, if enough people receive their flu vaccinations then the expansion of the influenza virus into the broader population is halted. Unlike the non-model approaches, modeling here may provide a statistical measurement of the probability of a halt to disease transmission with the goal of herd immunity. It does not guarantee the halt but raises the probability to an acceptable and practical level.[35]

A model may even produce no fruit at all. A model might become a component in a larger test and so provide a step toward a result but that result comes from the sequence and not the model. A model may also produce a result consistent with a theory and it may also exist only for the purpose of examination without the benefit of a precisely stated thesis. You can see this in weather forecasting, where the model of a rotation is shown as a part of a larger front, which is part of an even larger system.

Models also do not produce a deductive proof because a model is not a deductive test. Any conclusions drawn from a model is an inductive conclusion. It is based on repeated and varied tests against the model whereby a statistical conclusion is reached.

There are three ways in which models may be employed. (There is a fourth, but it follows in the next section because it has some unique characteristics that make it stand apart from these three.) The most common is *conceptual replication and emulation*. (This parallels what Craver called *experimental design*.) Engineers who construct soft models within a CAD/CAE environment and then operate these models entirely within the confines of a computer before anything physical is ever built commonly use this approach. The approach has value not only in the creation of a device but also in diagnosing problems at a later time. In both cases the same model and method is employed though it may operate with a different set of constraints depending on the purpose.

The second use is *explanatory*. Almost everyone has seen molecular models, toy car models, and other similar items. These models serve a simpler function. Instead of leading to a functional conclusion they instead lead to understanding. Some of these, like a molecular model, may be employed toward functional ends, but that use is not required, as they may stand alone.

The third use of models is *information evaluation*. (This is similar to Craver's *control* use of models.) In this situation the model is used to evaluate information and provide a solution to the nature and character of the information. Information is placed into the model in a suitable arrangement according to the assumption of the truth value of the model. Think of it as a type of filter for qualifying the data through an assumed process. The process may not be thought of as "true" but instead functional, suited to a need. With that filter one can collect and arrange results and qualify the data accordingly.

Of these three, evaluation, explanation, and emulation, two of them can produce nothing new. That is, their fruit tends to be abstract. Evaluation and explanation place data into a pre-determined systematic or framework, and these require certain presuppositions that define the character of the model's fruit. This leaves them begging the question and in doing so they also become unfalsifiable in the sense that we understand it with empirical testing. We will just have to find different methods for falsification.

The fruitfulness of any model raises questions regarding the demand that what is science must be fruitful. A predictive model produces nothing. It anticipates a result. One might alternatively think of an accurate prediction as a fruit. But it is often a one-time event and is generally unrepeatable, as with a weather forecast.

An evaluative model exists for analysis. Like insurance actuarial tables or data mining it exists to examine what is rather than the production of something new. The results may then be employed in a subsequent predictive model, but in and of themselves they amount to filters for organizing data.

An emulation model replicates something, but again does not actually produce a result. Only the mechanism/schema model is involved in creating an output that substantively matches that feature. That capacity can be thought of as a real fruit from the model, where the others lack this capacity.

Mechanism (Schema) Models

A close sibling to the afore-mentioned models, this next type of theory describes a mechanism[36] and its processes. It shares with modeling the need for repetition and emulation but becomes clearer as the mechanism model attempts to produce a measurable output through a more concrete process description, or schema. The schema produced by a mechanism theory builds a structure of higher order activities and their required lower order activities. These are built as dependencies, or gates, with a variety of and/or conditions and other conditions attached in their necessary relationships.

More simply, the mechanism approach can be seen as an extension of some basic model approaches. The difference here is that we are looking first at the structure or pattern, the scheme, of the material. These also tend to be close, if not physical implementations, of something like the principle being assessed.

An example of this is electronic circuit design. Though generally used in design, circuit emulation is frequently used for engineering and reverse-engineer processes and so creates a new (virtual) item that behaves as does the physical unit.

Another example of a mechanism theory is identification of process dependencies, frequently used in business process analysis (IDEF). The fruit is identification of process dependencies produces a finite state description of operations. In the case of IDEF the results are employed to improve processes.

In the physical and biological sciences the mechanism model is evident in molecular biology. We have all seen drawing of atoms and molecules, especially representations of DNA. These drawings represent the various gene combinations available for trait transmission.

Mechanistic representations such as this allow for functional experimentation. Some tests of chemical behaviors and the subsequent changes in molecular structure are treated as deductive tests. When you add an acid to a base the result will be neutralized to some degree and the chemical reaction is measurable and observable. But combine two parents' DNA and the results are as various. Some features are predictable (dogs do not give birth to lizards and vice versa), but colors, height, strength, and other features are not so discernable. At least not yet.

A mechanism may also attempt a physical reconstruction. An example of this might be an atomic collider, attempting to emulate, physically, something that is considered to have occurred eons ago in the early universe, or in some subatomic structure. The fruit of such an experiment may be the creation of a new element or the splitting of an existing element.

Model Falsification

The empirical movement of 19th and 20th centuries raised questions about knowledge. Hume, Kant, and others asked how we can know and how we might verify that what we know is true. Hume began with empiricism but had an issue with the reliability of induction. Kant attempted to resolve the question by (among other things) separating reason from experience. [37] He did not resolve all the issues of induction but instead opened the door to empiricism with the application of *a priori* language was employed in the development of principled rules and laws for nature. This application of language has allowed empiricists to construct theories that appear to be fixed natural laws but which are in fact only the conclusions of induction.

Then science changed. Einstein's abstractions about the nature of nature forced the development of theories that were far less specific in both structure and fruit. It was no longer sufficient for a scientist to manage chemicals in a laboratory and thus conclude something about the earth's atmosphere. The study of the atmosphere became a statistical concern with probabilistic results. So while a laboratory experiment might be shown to work or not work under various conditions and limitations on language these new fields of inquire were semantic and rarely if ever conclusive.

Let's go back to the laboratory experiment for a moment, though. When hydrogen is combined with oxygen the result will be both heat and water. But the theory "what do you get when combining hydrogen and oxygen" is broad. On the surface of the sun the results would be different. But obtaining a different result does not mean that the theory is false. It only means that the language used was too imprecise. The language required additional precision to limit the scope of reliable results.

If we add to this theory the (hypothetical) conditions "under 0.1 and 2.0 earth atmospheres of pressure" and "between -200 and +200 degrees Fahrenheit" we might get a condition where this always occurs. But even then there are potentially missing conditions and constraints. If we used a close container or a large volume of each chemical there might be a violent explosion which otherwise would not have occurred. The precision required for verification and falsification provides a higher sense of reliability to the results.[38] Falsification is less a true-false test to "conclusively establish a universal generalization"[39] but rather a challenge to precision and reliability.

Little in science is as certain as empiricists and certain evangelists of the field might claim. Science is now more in the business of models rather than managed observational analysis. Cleland is thus correct that "falsificationism cannot be used to justify the superiority of one science over another"[40] since the methods of the new sciences vary greatly from traditional empiricism.

One consequence for consideration is whether and how a model approach might be subjected to falsification. Since the structure of a model differs significantly from the structure of an empirical test it would seem to follow that falsification occurs differently. In our example of combining oxygen and hydrogen we reach precision and reliability by exclusion. That is, we refine the hypothesis and the test conditions. Those conditions where the experiment would more likely fail are added so that the remaining state would be one where the success of the test would reach a high level of reliability. The greater the precision that is added to constraints will result in a higher reliability in the results. Increased constraints reduce the opportunity for both false positives and false negatives.

Of the various models available, the predictive and explanatory (historical) models seem to have greater availability to some sort of falsification. The most common predictive model to most people would probably be the daily weather forecast. This type of model gains its accuracy through an additive approach. Since it attempts to discern effect from cause, greater the quantity of potential causes added on the front end leads to a greater precision in the predicted outcome. The results are not understood as specifically true or false unless of course one is depending upon a specific outcome as related to the weekly golf game. The goal is not to predict what will happen on every square foot of ground but to predict what might happen in different areas at approximate times. The forecast (as of last evening before writing this) for this morning was +8°F but the actual was +3°F. It was in specific error but in general a useful number. It was precise enough for the needs of a population.

Historical models are more difficult to falsify. The answer in the historical model is the question of the predictive model. That is, the content of an historical model is the data that is fed into a predictive model. The concern here is whether there is a full set of data and whether the model's prediction(s) are justifiable with respect to the data.

This problem occurs frequently in court cases where the prosecution does not have enough evidence for a conviction. The model cannot lead to a conclusion that one might reasonably doubt. If the defense is able to find the model's weak point then there is a good chance, with a persuaded jury or judge, for the defense to score a victory.

The predictive model is composed of causes and seeks an effect. The historical model is composed of effects and seeks a cause. The historical model is also limited by the insights of the theory posited. As Popper noted the assumptions taint the results.[41] Feyerabend took this further to suggest that much of science is therefore impossible[42]. In this same period theologian Cornelius Van Til raised the same suspicion but went further by suggesting that knowledge is tainted by the human condition and is as such suspect.[43]

One characteristic shared by these two general model constructs is that they depend upon a volume of information to obtain greater accuracy at the output. We can ask, for instance, what were the causes of the American Civil War? The resolution of the slavery issue is the first that comes to mind. But the result gets more specific and accurate when other recent events are added to the query. The question of states' rights (constitution vs. confederation) becomes significant though the system change (from the Articles of Confederation to the Constitution) was roughly 80 years prior as the Confederacy sought to remove the then constitutional form of government in favor of a more libertarian confederation. Theological differences played a small part on both sides with the South being in a (globally) anomalous position on this issue[44] set against the dominance of postmillennial theology in the North working to create a better world[45]. The British influence by way of William

Wilberforce also contributed greatly. A multitude of factors play into the cause of the conflict and identifying no single event or condition is sufficient to fully explain the situation. There may be a minimal set of core issues accompanied by a multitude of cascading activities, all to explain historical events. Oversimplification makes for bad history.

Under-determination is another problem for historical models. Much of the data necessary for a reasonable conclusion is frequently missing, lost forever. As a result what might have been is inferentially predicted to fill in the blanks. But these points may have even less support than the initial model hypothesis. This can be observed almost annually in the search for a common human ancestor.

Let's pursue an example of this: In a court case it is suggested that person A killed person B. The historical hypothesis is that "person A committed this act against person B" and the prosecution lines up the historical evidence: Person A was the only person in the room with person B. Nobody else entered the room or exited the room. Person A was the only person in position to commit this act.

But now the defense makes its case: Did the room have windows? Did anyone else have cause to commit this act? Why did person A not possess the weapon necessary to commit the act? A room with a window provides opportunity to another to commit the act. The lack of a weapon means that another person may have been the perpetrator. If person B had made a sufficient number of enemies in life then others would have had cause to commit the act. So the prosecution goes back and gathers additional information. The initial hypothesis was underdetermined. Falsification for the hypotheses does not show it false but it does show it insufficiently precise to account for the effect. Like the predictive model this is a data-accumulating approach to modeling.

When the prosecution does not have enough historical evidence the case is underdetermined. Prosecutors know this all too well. Prosecutions not pursued for "lack of evidence" are underdetermined situations.

Historical analysis often leads to false positives as well as false negatives. As noted, an incompetent defense attorney might allow a prosecutor to lead the jury to a false positive conclusion regarding the guilt of a defendant (the conclusion is "guilty" matching the hypothesis even though missing evidence would indicate otherwise). Likewise a false negative might occur where the guilty party goes free on account of an underdetermined prosecution.

These differ from the self-verifying results of Marxists and Freudians as noted by Popper. Conclusions reached by underdetermination are one thing, but conclusions that merely reflect the hypothesis always find cause for self-justification. This is not science.[46] A hypothesis may be correct or incorrect, reliable or unreliable. But without a richly determined model (the more data the better) to act as support it serves as nothing more than a hypothesis (the less data the weaker the theory). A theory which will not account for data, or which does so in an insufficient manner, is not one whose conclusions should to be taken seriously. It may be accurate but it is not precise.

The term falsification itself is a bit unreliable. Induction and abduction are about what is likely or frequently accurate. It may snow or rain today but it may not snow or rain at my house. Such is the nature of induction in science. The truthfulness or falsehood of these models comes by degree. It is not often a yes or no proposition.

This does not mean that there are no general laws. It only means that these laws are not quite so fixed and unbending as one might think. They are general laws; they are principles. If the general laws are understood as the conclusions of induction then we are free to modify them as additional evidence allows fuller determination. This approach avoids the skepticism of Kant, Hume, and Feyerabend, alternatively allowing for induction to reach not specific facts but working conclusions with respect to these two model-type theory structures.

Concerns and Consequences

It would be futile to seek a science that, per Feyerabend, is not tainted in some fashion by culture and history. We are not capable of removing ourselves to a position of pure objectivity in order to produce some result that is incorrigible. An alternative to this is to gain an understanding as to how our personal and social history might be affecting our work. We can at least ask ourselves what might be influencing our work and affecting to some degree our conclusions.

The world of scientific inquiry is far from monolithic when it comes to an awareness of the constraints needed for doing one's best work while also being aware of external influences. Though the rules and constraints of science have general agreement it does not take much reading to see that scientists are not philosophers and philosophers are not scientists. The result, it seems, is that scientists may become overly concerned about their conclusions and may at times dismiss the rules of doing proper work. Steve Eaton and Diederik Stapel represent real and extreme situations while *Krippendorf's Tribe* made light of the problem. Still the problem exists, I suspect more broadly than we might realize, while unnoticed and too often unchallenged.

Even when work is performed according to the standards for one's field it is not difficult to begin inserting assumptions into conclusions. That is, since we are all organized thinkers and work from a systematic knowledge of a field, forcing data to fit the system may happen without our awareness. This systematic issue may be either the framework into which the evidence fits or it may be the method used to interpret the information.

The latter can be seen in interpretation. Today there exists a verificationist tendency, a requirement for some identifiable outcome. Some, as we have seen, will reject any approach to science that does not produce fruit, but in practice will denounce only those that might include some influence outside of the observable and measurable while still working with those theoretical models that produce no fruit.

The insistence that the only meaningful statements are those that are verifiable is today known as *logical positivism* (alternatively known as *logical empiricism*). As noted earlier here are those who have understood that this is not the case so we have the theoretical sciences. Topics of current scientific conversation, as earlier noted with string theory and tachyon theory, exist entirely outside of what is physically testable for verifiable.

Contrary to this, some will say that things such as mathematical constructs are testable within themselves. That is, each can be validated in terms of being coherent, consistent, and complete. Just like computer software, the evaluation of these formulas is an internal evaluation. This is treated as a type of empiricism, per the ORV (the Once Received View), even though there is no measurable and reportable result. This seems to be the direction of Wolfram (*A New Kind of Science*) where all of reality might be reduced or deconstructed to a set of mathematical formulae.

Included in these assumptions is the idea that science is about facts and religion is about faith. This assumption sets up a false dilemma that leaves the appearance that there are only two choices. As we have seen, not all science is about facts. Though good science is about having a quality theory, some of a theory, even if fruitful, might even be completely false. What one considers a "fact" becomes more fluid than first thought, and "faith" (in the reasonable character of a theory) might have a great deal more weight than many have at first allowed.

Information is always subject to the interpretive process that the chosen methodology establishes. Any "truth" or "fact" information uncovered is filtered according to the model. Models themselves are the fruit of assumptions and presuppositions, and as such are also not neutral. A model may be fallacious even though the information analysis that it produces may be consistent with the model. Information is thus placed under the constraint of the language of the ORV, the structure of the model, or their function in the defined or discerned mechanism. Going further, even these functional assumptions regarding methodology are controlled by other higher-level assumptions about the nature of reality and other questions. Thus science, even the natural sciences, does not exist apart from the interpretive process. It appears that objectivity in science is to some degree an illusion.

II. On Science and Origins

Origins and scientific inquiry are, in the eyes of many, Christianity's greatest apologetic challenge of the past two centuries.[47] The modern theories on origins have provided a mechanism for pitting naturalism and a materialistic view of the universe against theology and the concerns of the church. Though various theories of science had been floated long before Charles Darwin presented his theory none of them were adequately coherent systems and so faded into relative oblivion. But Darwin presented a fuller system which, though still rife with errors and unanswered questions, set forth the desired principles in such a way that general acceptance seemed almost certain.

The issue that we are presented with is not evolution *per se*, but the package that naturalism and evolution together bring to the table. Naturalism is brought along as the assumption, the supposed condition in which which exists. It is the assumption of a wholly material universe that is closed from divine intervention. It is taught that evolution is the product of natural events and conditions and that there is nothing else in the universe that might violate the assumption of metaphysical naturalism. It is here that we may challenge that pairing of evolution as a process when it arrives with naturalism as its assumption.

It is probably best to avoid the temptation to say that a naturalistic theory of origins is not science. It is.[48] It is the accompanying *naturalism* that is not science. Naturalism is an assumption, a presupposition. It is philosophy. It is a *why* question whereas science is a *what and how* question. It is a metaphysical position. Any subsequent suggestion that naturalism is proven by evolution is a circular argument. Naturalistic evolution is employed to serve as the evidential apologetic for naturalism. The mechanism of evolution is our first concern.

This section is composed of two parts. The first is to cover the general principle of evolution. This is a summary, a brief snapshot, of what evolution is supposed to be – how it is understood to be found working, so the statements made tend to be broad and encompassing.

The second part is how evolution is said to work. Here the discussion gets more specific as the discussion moves to the *engines* of evolution. These are the processes that are understood to drive the natural world. These might be thought of as the idealized models of how evolution functions.

Some Characteristics of Naturalistic Origins

The goal here is to get just a snapshot of the major views from within the Darwinian camp. These snapshots come with some of the characteristics that one might expect to find. Some of these characteristics are necessary to maintain the model while others hold significance because of their implications.

In the natural sciences there are many who hold a variety of views about the subject yet all are regarded as scientists. Michael Behe and Thomas Nagel, for example, have not endeared themselves to the naturalistic community even though they remain generally committed to some sort of naturalistic system. Behe is a scientist and an adherent to intelligent design and Nagel is a philosopher. Their approach might be classified as non-Darwinian. Likewise there are some minor views, such as Thomas Wolfram's mathematical and computational proposal, the suggestion that complexity can result from some sort of recursive system.

Within the natural sciences the term evolution is generally used in the systematic Darwinian sense (common descent with modification, along with everything that cascades from that principle) but at the same time may be loaded multiple meanings. Though the concern here is naturalistic biological evolution and the development of species the term also has been coined for other uses. At times it is used in a vague and generalized sense to describe how anything develops. One might read, for example, of the evolution of an idea or a business. Such statements make no appeal to any beginning or end to the purpose or function of such an idea. It is reduced to an appeal to some indefinable perception of change. With it comes the hint that such change is always good – especially when it is presented as being as a change for your own betterment. One might also hear the term used sociologically regarding, for example, the evolution of group behavior. This may encompass various areas such as religious principles or cultural trade development. Again, these terms are rather vague.

Employing the term outside of biological science is somewhat controversial. It also appears that the term *Darwinism* may be nearly archaic in its usage. *Evolution*, especially when used in this more general sense, seems the more common expression. It encompasses the fields of history, fossils, and molecular biology whereas Darwin's material was observation along with estimations and calculations based on his observations.

Let's begin with a broad definition of evolution. It is the proposal that *the presence of all species on the earth can be explained in terms of common descent.* That is the core principle of Darwin and all that followed him. It has developed into a richer model that attempts to correlate historical information into a systematic explanation of life. But to be more precise we should say that there are several alternative models that attempt to systematize the data. Though they all are identical at the highest level (common descent) once one goes beyond that point their frameworks diverge significantly.

Darwin's concept is the broadest way to define evolution. Evolution is *common descent.* That is the core definition of all of their work. What remains is how all that is explained. For Darwin[49], "I am convinced that Natural Selection has been the main but not exclusive means of modification," says that behaviors with regard to reproduction and adaptation lead to new species development. That's the basic principle at work.

Ernst Mayr

Ernst Mayr was a well-regarded mid-20th century scientist who has given us some insight into the modern development of evolutionary theory. According to Ernst Mayr, Darwinism remains (at the time of his writing) at the core of evolutionary theory. He says:

> Darwin also saw clearly that there are two aspects to evolution. One is the "upward" movement of a phyletic lineage, its gradual change from an ancestral to a derived condition. This is referred to as anagenesis. The other consists of the splitting of evolutionary lineages or, more broadly, of the origin of new branches (clades) of the phylogenetic tree. This process of the origin of biodiversity is called cladogenesis. It always begins with an event of speciation, but the new clade may become, in time, an important branch of the phylogenetic tree by diverging increasingly from the ancestral type. [50]

He also uses the term evolution in that generalized sense, and I think in a consistent fashion, and applies it to the rest of the universe when he says:

> Almost everything in the inanimate universe is also evolving, that is, it is changing in a distinctly directional sequence. [51]

Mayr works with the assumption that the evolutionary changes that occur are, on the broader scale, taking place in a beneficial direction. This is called directionality. This pretext appears again as he works to reconcile chance and necessity in evolutionary processes:

> Many authors seem to have a problem in comprehending the virtually simultaneous actions of two seemingly opposing causations, chance and necessity. But this is precisely the power of the Darwinian process. [52]

The power of the Darwinian process looks forward to the process providing generally beneficial results so that humans, and later perhaps even a greater being, a better world. In response to this we ask the question of whether or not humans continue to advance?

> Modern humans, by contrast, constitute a mass society and there is no indication of any natural selection for superior genotypes that would permit the rise of the human species above its present capacities. [53]

He also accepts earth's, and thus humanity's, apparent uniqueness in the universe.

> Even if something parallel to the origin of human intelligence should indeed have happened somewhere in the infinite universe, the chance that we would be able to communicate with it must be considered as zero. Yes, for all practical purposes, man is alone. [54]

The uniqueness of humans in the universe is not specifically an evolutionary question. But it does raise questions regarding various proposals that would suggest that the earth's evolutionary process is happening elsewhere in the universe and is even a frequent occurrence. If it is not a frequent occurrence then some understandings of the universe and the assumptions being drawn from those suggestions might require some reevaluation.

Stephen Jay Gould

Stephen Jay Gould entered the forefront of these discussions in the latter half of the 20th century. His (along with Niles Eldredge) tome, *The Structure of Evolutionary Theory*, has been the bible for many studying evolutionary biology. In one significant volume he overviews the science and philosophy of evolution. His view of Darwin appears not significantly different from Mayr. His work is in fact an expansion and a corrective of

Mayr.

He views Darwin as a philosopher as well as a scientist[55], a fact that should never be missed[56].

> Many scientists fail to recognize that all mental activity must occur in social contexts, and that a variety of cultural influences must therefore impact all scientific work. Those who do note the necessary link usually view cultural embeddedness as an invariably negative component of inquiry – a set of biases that can only distort scientific conclusions, and that should be identified for combat. ... The origin of Darwin's concept of natural selection provides my favorite example of cultural context as a promoter.

His definition of "evolution" appears as an expansion on, clarification of, and built upon the foundation of Darwinism. He is not really Darwinist in the adaptationist sense (like Mayr he is a neo-Darwinist) but employs the term in a broad sense so as to define modern scientific theory (relatively modern, that is, as of the late 20th century) within the scope of what Darwin began. The same word has taken on a slightly different meaning given this new context. Gould thus uses the current conventional terms "synthesis" and "consensus" to describe the movement and its adherents despite disagreement on the details. [57]

Donald R. Prothero

Dr. Prothero's definition of evolution seems to be clarified in two expressions. The first of these is his paleontological proposal – fossils. From this vantage point he appeals to a proposed mechanism behind those fossilized creatures. He starts with a discussion of abiogenesis[58] and then proceeds to a type of adaptationism. He describes something that appears to be more like cohabitationism. Appealing to the efforts of Margulis and others he suggests that small one-celled creatures merged with each other to form bigger creatures. Those little things adapted by merging to form a bigger thing.[59]

A significant point in his work is his documentation of the change in classification and relationship method that is used by scientist. For the most part the old phylogenic (fossil-based) tree is gone (at least as the standard reference), and the reason it is gone is because it cannot be filled in with enough fossil evidence. That tree, whose intention was to make the historical links from the past to the present, has been replaced by the cladistic tree. This tree is used to attempt to answer some difficult questions. But as he says "the biggest sticking point has been the concept of ancestry." [60]

> The biggest sticking point is the concept of ancestry. We tend to use the term ancestor to describe certain fossils, but we must be careful when making that statement. If we want to be rigorous and stick to testable hypotheses, it is hard to support the statement that "this particular fossil is the ancestor of all later fossils of its group" because we usually can't test that hypothesis. Because the fossil record is so incomplete, it is highly unlikely that any particular fossil in our collections is the remains of the actual ancestor of another taxon[61].

He also says

> But there's another reason why cladists avoid the concept of ancestry. To be a true ancestor, the fossil must have nothing but shared primitive characters relative to its descendants. If it has any derived feature not found in a descendant, it cannot be an ancestor. [62]

A recently developed method known as molecular phylogeny has allows this concept to be developed further. Advances in DNA research have allowed the creation of cladistic trees. These are employed to establish the genetic relationship between various kingdoms of creatures.

Or more simply, the fossil trees attempted to build a visible hierarchy (which could not be attained) whereas the cladistics tree builds on genetic relationships. It's a trade-off, minimizing the visual similarities of the fossil trees and maximizing the genetic relationships. The fossil trees could not establish broad relationships and the genetic trees have some difficulty with ancestry.

One weakness in Dr. Prothero's view is how he defines science. This seems more of a distraction and misuse of language but is also represents a structural problem in his argument. He sees evaluating the scientific process as one of deductive reasoning[63]. I find this surprising since, as a zoologist he would or should recognize historical models as either inductive or abductive as he searches for the most plausible answers to his questions. Deductive reasoning, on the other hand, says that the conclusion follows from the premises. One simply may not do this with arguments from historical evidence. One does not reach a deductive conclusion from an inductive (or abductive) argument.

Of course what Dr. Prothero is talking about is something more industry-specific. He uses the term "hypothetico-deductive" to describe the approach. Despite the name it is not the deductive reasoning implied by the name. It is more akin to the loose use of the term in a Sherlock Holmes adventure. It might be better phrased "concluding from the hypothesis." It amounts to a re-defining, or perhaps an archaic use, of the term deduction so that the conclusions seem to gain greater acceptance by the reader. After all, deductive conclusions would hold more weight than inductive ones.

What does it mean? It is a linguistic matter, really. The goal of the language is to avoid the problem of generalized statements so that the scientific process might be more precise. As he describes it, it's about hypothesis support rather than making generalizations, similar to what we did with our language concerns with defining science. It is also more about how the conclusion is stated and has nothing to do with the process itself and that seems to be where the rub lies. The processes remain inductive or abductive. The theory *structure* never changes. What changes is the language used when reporting results.

Merriam-Webster defines deductive like this:

> relating to, being, or making use of the method of proposing hypotheses and testing their acceptability or falsity by determining whether their logical consequences are consistent with observed data[64]

Another application of this method appeals to the traditional scientific method and at the same time harkens to the approach of the empiricism of past decades.

> From the hypothesis, the researcher must generate some initial predictions, which can be proved, or disproved, by the experimental process. These predictions must be inherently testable for the hypothetico-deductive method to be a valid process.

> For example, trying to test the hypothesis that God exists would be difficult, because there is no scientific way to evaluate it.[65]

In the end it amounts to nothing more than an alternative name for the traditional scientific method but with a language change to include Popper's falsification along with the constraint that the theory be precise. This demand is helpful to better work even though the tendency toward empiricism and assumed deductive conclusions would be in error.

Dr. Prothero also appears to be open as to the engine of evolutionary change. Zoology and paleontology do not depend upon Darwin's gradualism or Mayr's synthesis or Gould's variation of the synthesis, punctuated equilibrium. Yet he holds a biochemical model regarding origins that would be difficult to model and replicate.

Richard Dawkins

Dawkins' view of evolution gives us yet another perspective on the matter. He expressed his perspective that Darwin's assumptions of gradualism and adaptation appear to be the most suitable. In this he sees evolution not as a matter of the transmission of community traits through genetics, but changes that belong first to the individual. That is the selfish gene he wrote about.[66]

Dawkins' perspective on the evolutionary process is probably best summarized as closer to Darwinian adaptationism. It is the individual who survives and adapts and whose genes develop into something of greater capacity. In this he differs with Ernst Mayr who viewed the process as one of community rather than individuals.[67]

The Driving Forces of Evolution

The selected scientists have given us a broad picture of the evolutionary process. But this breadth might prove to be a distraction if the function is forgotten. These theory structures all have a plan of action – a framework for understanding how things appear to have worked in the past and how they appear to work today. They attempt to explain the processes of the past and present.

Common descent is the assumed fruit of evolution and not its engine. It is not what drives evolution. The principles of descent are the outcome of the engine's function. It is what happens as a result of the work of whatever engine is operating behind it. Though evolution as a process may be treated as the engine, the real engine is whatever mechanism drives common descent. The question here is not as much *what* happened but *how* it happened. That is the function of each model and that is what makes each one distinct.

Adaptationism: Darwin knew little or nothing of genetics as we know it today. As he understood the world around him traits that a creature developed could change through the process of adaptation. His primary engine was behavior. (His language indicates that he knew that there was something else working behind the scenes but he couldn't identify it.) For some this behavior was fitness and strength. For others it was sex selection, especially for humans.

The first synthesis: After Darwin's theory was evaluated and new information came to light there came an adaptation to his theory. From Mendel's genetics theory there grew a group that understood that genetics was the driving force. Mendel's was an experimental approach that showed traits being passed on in the hybridization process. The method remains popular in certain fields of study.[68]

With the newly-developed science of genetics (though still pre-DNA-knowledge) there arose the concept of mutation. But unlike Darwin this was not behavior-driven mutation but genetic inheritance-based, random mutation. Genes alter features and new features mean a new context for living.

Of course one should not think that the simplicity of Mendel's pea variations was applied with naïve simplicity to common descent. Things were developing rapidly in the field. Julian Huxley, an evolutionary biologist and eugenicist, again at that time not knowing about DNA, described the feature-bearing chromosomes and centrosomes which transmit features to both reproduced cells via mitosis and to offspring.[69] For Huxley and others of the genetics-driven solution to how evolution works, genetics is everything. It is genetics that prevents yellow mice from reproducing[70] and which created the Hapsburg Lip[71]. His conclusion of all this is to "set aside on-Mendelian heredity"[72] as he finds no other source for features than genes.

This approach is far from dead and a variety of alternative configurations can be found in this arena. For instance, Harmit Singh Malik is "an evolutionary geneticist at Fred Hutchinson Cancer Research Center who studies genetic conflict – the competition between genes and proteins with opposing functions that drives evolutionary change."[73] This is, of course a very basic statement regarding the process. But the principle and question remain the same – what drives evolution? For the early neo-Darwinists it was genetics before behavior and adaptation.

It was not long until people began asking if genetic change alone were enough to provide the change. Perhaps there is more that needed to be developed. So now we get past the individual and into group evolutionary dynamics.

Neo-Darwinism, the second synthesis: Now neo-Darwinism comes into full form. Ernst Mayr's (among others') understanding is that the engine of evolution is accomplished through accumulated genetic changes within a community. This is the synthetic theory – the synthesis of genetics and adaptation.[74]

> The proponents of the synthetic theory maintain that all
> evolution is due to the accumulation of small genetic
> changes, guided by natural selection, and that transspecific
> evolution is nothing but an extrapolation and magnification
> of the events that take place within populations and species.

Mayr's divergence from classic Darwinism is clear here. For Darwin
the process was about natural selection aided by other forces. It's not that
there were no other forces, but he was not fully aware of them and the
capacity to describe them had not yet matured. So he set his priority where
he could. For Mayr natural selection is the context in which the genetic
changes create the real change.

Group genetics became the new driving force rather than individual
genetics. One analogy that might work here is the higher frequency of red
hair in the Germanic (northern European) population.[75] This genetic trait
or mutation created a feature that has not as yet become a general
population trait. Should all the right people mix then it would be possible
for the trait to be seen more widely, though at this time it is a minority
feature. We all know of features that identify groups – eye, hair, nose, etc.,
are often employed to identify something about a person's genetic
heritage. Red hair, for instance, among those of German and (by
migration) Irish descent.

Dawkins argued against Mayr on this. He noted how genes behave and
that, as he sees things, the process depends upon one individual to change.
The change has to start somewhere and at some point. It is at this point
where the selfish gene chooses its own survival above all else.[76]

Gould also takes a position as a correction to Mayr though he goes a
different direction. Gould and Eldridge reformulated the synthesis. Now
we get a change to one of Darwin's significant features — gradualism.
Gould and Eldridge proposed periodic rapid changes over against slow,
indiscernible changes. "All scholars have noted the centrality of
gradualism, both in the ontogeny (Gruber and Barrett, 1974) and logic
(Mayr, 1991) of Darwin's thought."[77]

Now a key component is up for discussion. They, Gould and Eldridge,
concluded that species-level changes might occur rather abruptly[78]. They
acknowledged the "discordance between theoretical expectations and
actual observation" and the issues created by that situation. From their
perspective the process of speciation must happen quickly in order to
accommodate the fossil record and account for the apparent lacks in the

data. This is to them the weakness of a purely paleontological[79] approach. Working from the fossil data is not enough.

James Shapiro has identified this shift in neo-Darwinian thought in perhaps more general terms.

> Conventional evolutionary theory made the simplifying assumption that inherited novelty was the result of chance or accident. Darwin theorized that adaptive change resulted from natural selection applied to countless random small changes over long periods of time. … His neo-Darwinist followers took the same kind of black-box approach in the pre-DNA era by declaring all genetic change to be accidental and random with respect to biological function or need. With the discovery of DNA as a hereditary storage medium in the 1940s and early 1950s, the accidental view of change received a molecular interpretation as arising from inevitable errors in the replication process. As any professional and popular press articles attest, the accidental, stochastic nature of mutations is still the prevailing and widely accepted wisdom on the subject.

> In the context of earlier ideological debates about evolution, this insistence on randomness and accident is not surprising. It springs from a determination in the 19th and 20th centuries by biologists to reject the role of a supernatural agent in religious accounts of how diverse living organisms originated. While that determination fits with the naturalistic boundaries of science, the continued insistence on the random nature of genetic change by scientist should be surprising for one simple reason: empirical studies of the mutational process have inevitably discovered patters, environmental influences and specific biological activities at the roots of genetic structures and altered DNA sequences. The perceived need to reject supernatural intervention unfortunately led the pioneers of evolutionary theory to erect an a priori philosophical distinction between the "blind" process of hereditary variation and all other adaptive functions. But the capacity to change is itself adaptive. Over time, conditions inevitably change, and the organisms that can best acquire novel inherited functions have the greatest potential to survive. The capacity of living organisms to alter their own heredity

is undeniable. Our current ideas about evolution have to incorporate this basic fact of life.[80]

Shapiro has done two things here. First he makes his return appeal to Darwin but with the caveat that Darwin's theory, and those of his successors, was influenced by the times, was also accompanied by a subjective component. These factors left the world with a flawed theory. Then he takes the argument further and introduces the idea that DNA is the driving force behind evolution. In this it is neither random nor blind. By doing so he escapes the potential problem of a soft saltationism and other issues that accompanied random and blind change. He also accomplished something quite important: The drive to survive has something behind it that moves it. DNA is not only the source for life's definition. It is the new defining force of life.

Shapiro's principle is a "cell-guided"[81] model that allows for rapid genetic and feature changes reminiscent of punctuated equilibrium with a mix of selfish gene theory. His model represents a clear rejection of "the simplifying assumptions made in the 19th and early 20th centuries" which he says are "plainly wrong."[82] "It requires great faith to believe that a process of random, accidental genome change could" … "duplicate and modify increasingly complex genomic constructs." "Indeed, as many biologists have argued since the 19th Century, random changes would overwhelmingly tend to degrade intricately organized systems rather than adapt them to new functions."[83]

These changes are at one point separated from the external cognitive influences of any creator[84] but that does not mean that there is no cognition involved. Shapiro identifies the content of the cells and DNA as a cognitive, direction-setting component. "Life requires cognition at all levels."[85] Recognizing this information as real information rather than random data introduces an clear teleology.

III. How (Some) Scientists Argue

Here we ask questions about how the apologists for naturalistic evolutionist deal with evidence and presuppositions and their influence on conclusions. This is where we begin to see some of the issues with how arguments are made.

There are several points to cover in this section. The methods used by naturalistic evolutionists in arguing for their conclusions come with multiple goals. Their core apologetic, their offensive, is the use of evidence. It is in the use of evidence where they seek to establish the theory as sound. At times the evidence may appear overwhelming to the reader or student. Such a mass of information is challenging to consider. But the important thing to remember is that evidence always comes with a context, a set of presuppositions and assumptions, which at times leaves these arguments vulnerable.

Naturalistic evolution also has a defensive apologetic, as all systems recognize the need to defend against attacks. In this case it is also against qualifications to the evidence that might unseat the chosen theory from its dominant position. For instance, there have been notable atheists who, while accepting a naturalistic theory of origins at least in principle, likewise reject Darwin's model as insufficient. There are questions that concern them. It is against their criticisms that the defensive apologetic is used internally within the scientific community.

An example of this is the description of "mind" from a purely materialistic perspective. Cognitive scientists such as Jerry Fodor might raise a question that goes beyond a purely materialistic approach. The materialistic response is to argue from the systematic. That is, they argue from their materialism. That being, nothing other than matter and energy can be accounted for. Therefore, to the materialist, the mind is nothing more than firing neurons. So instead of answering the objection the data is reframed according to the materialistic systematic.

External criticisms of naturalistic evolution also come from the world of theism. Naturalistic evolution demands that there is no deity behind the natural world. Any insistence on the necessity of a creator is automatically a target since theism and naturalistic materialism sit at odds. The nature of these arguments and defenses provides some insight into how the whole of the argument is constructed.

Another consideration is the character of their arguments. Are the arguments based only on the sufficiency of the information available? Or is there some tendency either to appeal to presuppositions or to demean the critic? The use of evidence is, of course, a controversial one. The weight that should be assigned to various pieces of evidence is always up for discussion. Likewise without some capacity to identify one's prejudices the evidence itself might be seen as tainted. Lacking this capacity seems to be the place where unnecessary slurs find their way into the conversation.

Fortunately not all scientists engage in unnecessary slurs. James Shapiro's book has none of that, and for that he has my respect. His interest is the science and with it some of the related philosophical concerns. While I find his science fascinating (especially how he manages to reconcile Gould & Eldridge with Dawkins) it is his philosophical system that seems lacking, as noted earlier.

The Positive Apologetic

The Presentation of Evidence

The presentation of evidence for naturalistic evolution almost always begins with the fossil record. This is foundational to the discussion. On the surface it appears to be a simple matter. Just dig up fossils, identify their age according to the surrounding material, and let the evidence speak for itself.

Still there remain some variations in this understanding. The concept of cladistics is the grouping of biological beings in terms of features – shared characteristics. The method is in regular use though it has some issues. Dr. Prothero clarified the issue.

> The biggest sticking point is the concept of ancestry. We tend to use the term ancestor to describe certain fossils, but we must be careful when making that statement. If we want to be rigorous and stick to testable hypotheses, it is hard to support the statement that "this particular fossil is the ancestor of all later fossils of its group" because we usually can't test that hypothesis. Because the fossil record is so incomplete, it is highly unlikely that any particular fossil in our collections is the remains of the actual ancestor of another taxon (Schaeffer et al. 1972; Engelmann and Wiley 1977).

> But there's another reason why cladists avoid the concept of ancestry. To be a true ancestor, the fossil must have nothing but shared primitive characters relative to its

descendants. If it has any derived feature not found in a descendant, it cannot be an ancestor.[86]

This has developed, as Dr. Prothero points out, in a change as to how cladists view information today compared with how they viewed data, say 40 years ago[87]. As I understand Dr. Prothero the current understanding of ancestry is to abstract the principle. Instead of attempting to fill in the complete fossil record with as many (let's call them "immediate") ancestral artifacts as can be found in an attempt to build a complete picture, the trend now is to observe and classify according to greater similarity. Doing this allows for classification by inference and predicting what might fill in the intermediate positions. This approach serves to refine a model without having to collect one hundred percent of the data the earlier approach to ancestry would have required.

The Imperfection Argument

An effective component of any offensive apologetic is to disarm one's opponent. By defusing the opponent's arguments one's position is left in a stronger position. Along this line, one of the more interesting discussions comes about when naturalist questions the character of creation. Among these arguments is the imperfection argument. This argument looks at the less-than-optimum nature of biological life and suggests such imperfections serve as defeaters for creation models. For example, regarding human anatomy.

> We might ask those same biologists, however, to explain why they think evolution is true. This is a task they face regularly, if only for pedagogical reasons. Consider a well-known example. The giant panda (Ailuropoda melanoleuca) possesses a pseudothumb built from the radial sesamoid, a wrist bone. The panda uses this structure, somewhat clumsily in the eyes of certain observers, to manipulate its main food, bamboo. Odd structures like the panda's pseudothumb, argues Stephen Jay Gould, "are the primary proofs that evolution has occurred" (1991, p. 61), for
>
> > If God had designed a beautiful machine to reflect his wisdom and power, surely he would not have used a collection of parts generally fashioned for other purposes .... Odd arrangements and funny solutions are the proof of evolution - paths that a sensible God would never tread but that a natural process, constrained by history,

follows perforce (1980, pp. 20-21).

This passage, from Gould's essay on the panda's thumb, is an instance of what is frequently called the imperfection argument for evolution. God is an optimizing creator. This structure, and hence, organism, is imperfect. Therefore this organism evolved.[88]

One response to the argument is to present this as the naturalistic version of the "God of the gaps". This argument, when used by the theist, declares that the unanswerable is of "God" as so dismisses the unanswerable. In the case of the imperfection argument's conclusion of "imperfect, therefore evolved" would meet that criteria. Or, we might state it this way: If the structure of the "God did it" is fallacious (a point with which I agree) then so is "imperfect, therefore evolved" equally a fallacy. The argument is the same because the structure of the argument is the same. Neither appeals to either evidence or even an appropriate agnosticism about the unknown.

The Negative Apologetic
External Criticisms

A negative apologetic is a defensive one. Every position, it seems, requires both an offensive (positive) and a defensive (negative) apologetic. How naturalistic evolutionists defend their position against external criticism comes in several forms. There are some non-Darwinists who accept naturalistic origins in general but not the Darwinian models. These persons are treated as outsiders in the scientific community. As a result their criticisms raise serious questions that often go unanswered or are simply dismissed (as in the question of mind as noted previously).

Some external criticisms also come from theists who challenge the assumptions and presuppositions and who challenge the results and fruit of the theory. Why, for instance, is it necessary to always couple naturalism (the belief that there is no creator) with a study of the natural world? Does doing so amount to begging the question on the issue? (That is, does it not beg the question to couple the presuppositions with the evidence?) And what of ethics? What does the inability of naturalism to work with ethical standards and morality mean to the world?

Non-Darwinism

One might think that the modern Darwinist system is the only way of looking at evolution. Darwin was, after all, the person who provided today's dominant mechanism – adaptation with sex selection – to explain the origin and development of species. But in the minds of many these criticisms of Darwinism, once it had become accepted and entrenched, amount to a rejection of The Truth, The Orthodoxy Science to Which Nobody May Disagree. Suggest something else and the result will be more than just a strong challenge. There are no alternatives. Barzun, a philosopher in the mid 20th century, noted this:

> So true is this that the ordinary educated man of today sees no third choice between the "scientific ideas" of the late nineteenth century and the "obscurantism and superstition of the Middle Ages." One can imagine him saying" "You are not a Darwinist? – You must be a Fundamentalist." "Not a believer in economic causation? – You must be a mystical Tory." "Not a materialist? – You must be an idealist." The implication is that if you are all of the latter things you must be on the side of ignorance, folly, and "reaction." And since these are justly dreaded evils, any critique of scientific materialism must be an attack on right reason.[89]

Things have not changed since Barzun wrote this several decades ago. He also noted one of the significant criticisms of Darwin's time by Samuel Butler. It was his conjecture that change produced fitness for survival rather than strength and survival producing change. Barzun says, "As Butler put it, 'To me it seems that the "Origin of Variations," whatever it is, is the only true "Origin of Species."'"[90]

Butler also raised this concern about the mind as an entity that could not be accounted for by such a mechanical view of life, as the mind is apparently not a simple mechanical entity. Jerry Fodor and Thomas Nagel have raised this same criticism in more recent times.[91]

Butler's notion made society something other than a Coliseum where human beasts strive with one another in total darkness. Mind, feelings, ethic, art – all these things once again became real, instead of being the dreams of automata, accompanying the physico-chemical changes called digestion, respiration, reproduction, and death. [92] This same criticism of Darwin's mechanical world has persisted, most recently in Thomas Nagel's 2012 "Mind and Cosmos" where he decries the loss of "mind" in Darwin's materialistic model. For a century and a half the question has gone without an adequate answer.

Assumptions

One of the principles of evolutionary theory is the assumption that any species, plant or animal, must adapt, must change, must improve or go extinct. Julian Huxley stated the problem this way:

> A very fine and curious issue to which we shall later direct the reader's attention is that of Orthogenesis. It is alleged that in a certain number of cases species, even though fairly well adapted to their conditions, and without experiencing any change of conditions, have by virtue of a sort of inner drive, an innate destiny of the species, gone through considerable evolutionary change. Professor Henry Fairfield Osborn finds this drive convincingly displayed in various fossil series, such as the horses, camels, and titanotheres. Of course, where the changes produced by the orthogenesis have been disadvantageous, Natural Selection has at last arrested the drive and extinguished the line. But the supposition that the new form, ne structure or other characteristic is not advantageous or is insufficiently advantageous to have "survival value" is difficult to establish.[93]

This idea of directionality is one of naturalistic evolution's weakest positions. With it comes an assumption about the character of the material world, that some unknown and unidentifiable principle is at work to move species to improve their lot has haunted evolutionary theory for centuries. Lamarck had his "use" based theory to explain how the giraffe's neck got longer (by using it, by reaching into higher branches generation after generation), thus improving the odds for survival. That argument did not work. Reaching further does no more to change a life's form than does cutting off the tails of mice cause the tails to disappear over time.

Then there is the corollary of non-use. Why would an organ disappear through non-use by Darwinian process if it did not, and could not, come into existence through the non-Darwinian, Lamarckian process of use? Jerry Coyne appeals to the non-use scenario when he discusses flight and predation in the Southern hemisphere.

> Let's take wings again. Obviously, there are many advantages to having wings, advantages shared by the flying ancestors of flightless birds. So why did some species lose their ability to fly? We're not absolutely sure, but we do

have some powerful clues. Most of the birds that evolved flightlessness did so on islands – the extinct dodo on Mauritius, the Hawaiian rail, the kakapo and kiwi in New Zealand, and many flightless birds named after the islands they inhabit (the Samoan wood rail, the Gough Island moorhen, the Auckland Island teal, and so on). ...

The long and short of it is this: flight is metabolically expensive, using up a lot of energy that could otherwise be diverted to reproduction. If you're flying mainly to stay away from predators, but predators are often missing on islands, or if food is readily obtained on the ground, as it can be on islands (which often lack many trees), then why do you need fully functioning wings? In such a situation, birds with reduced wings would have a reproductive advantage, and natural selection could favor flightlessness. Also, wings are large appendages that are easily inured. If they're unnecessary, you can avoid injury by reducing them. In both situations, selection would directly favor mutations that led to progressively smaller wings, resulting in an inability to fly.[94]

This criticism should be understood as a criticism of inconsistency. On the one hand the Darwinian model is the standard and then, when convenient, the Lamarckian model of change is taken as the answer. This inconsistency serves as the only practical way to answer the directionality question.

The Presentation of Data

Just one example is provided, but it is one where the explanatory model of the evolutionist has proven less than adequate. Peter and Rosemary Grant provide an example for model falsification in their discussion of Darwin's finch evolution:

Even if a model fails to fit the data perfectly, it is useful to describe evolutionary branching with a metaphor in mind, because the confrontation between data and metaphor encourages the posing of sharp questions. **Being forced to fit data to an idealized concept may lead to new insights and revised idealizations.** For example, as we discuss in more detail later, thinking about the loss of lower branches in evolutionary trees forces us to consider past extinctions and the contribution of those losses to the current form of a tree.[95] (bold emphasis mine)

The Grants here do one thing quite valuable and at the same time expose a shortcoming. To "revise idealizations" is to correct errors in the model. That is just good science. This is, in principle, what the non-Darwinists are demanding. But how to go about it – by shoe-horning the data into the model instead of re-evaluating the model itself – is to be committed to the systematic rather than the information. The problem is that the idealized model for them comes ahead of the working model that would (or should) appropriately accommodates the data. The two should be in sync, with the idealization being merely a generalization of the working model. This seems not to be the case.

What happens when an scientific model makes a false prediction? Should it be discarded as false? Should its error be glossed over? Or should we just talk over the issue and put on a mask of certainty.

Randolph Schmid, AP Science Writer, recently presented us with one of these situations in his article "'Game-changer' in evolution from S. African bones." In this article Darryl DeRuiter of Texas A&M says of this find

> This is what evolutionary theory would predict, this mixture of Australopithecine and Homo," DeRuiter said. "It's strong confirmation of evolutionary theory.[96]

That is evolutionary modeling as an explanatory structure and the term "predict" here has nothing to do with the future but is about anticipating what evidence might fill in the blanks. The question here is — is this correct? Does this represent the reality of the evolutionary predictions? The article goes on to say that

> For example, in previous definitions of our genus, the leading edge in the emergence of Homo has been brain enlargement. The sediba bones show, however, that reorganization of the brain and pelvis typically connected with the evolution of Homo need not have involved brain enlargement," he noted.

Now we appear to be approaching some serious confusion. Were the older models that supposedly made an alternative prediction, and which were presented as settled science, wrong? How wrong were they? Might this prediction be in error? And if so, to what degree?

"This is what evolutionary theory would predict, this mixture of Australopithecine and Homo," DeRuiter said. "It's strong confirmation of evolutionary theory." Yet what "need not have involved" sits in a bit of contradiction. What was assumed previously is now a contingent matter. Apologists would do well to note when arguments are at times modified for convenience sake without an acknowledgment of error. This problem leads into the next concern.

Arguing from the Model

To argue from the model (that is, the systematic, the *organization* of and imposed on the evidence) is to ignore the evidence. That characteristic was part of the Grants' error just noted. Jerry Coyne gives us an example of arguing from a subsequent hypothesis instead of from evidence. He says in a note:

> 4. It's likely that ancestral mammals retained their adult testes in the abdomen (some mammals, like the platypus and elephant, still do), which makes us ask why evolution favored the movement of testes into an easily injured position outside the body. We don't yet know the answer, but a clue is that the enzymes involved in making sperm simply don't function well at core body temperature (that's why doctors tell potential fathers to avoid warm baths before sex). It's possible that as warm-bloodedness evolved in mammals, the testes of some groups were forced to descend to remain cool. But perhaps external testes evolved for other reasons, and the enzymes involved in making sperm simply lost their ability to function at higher temperatures.[97]

What Dr. Coyne has done is quite simple: He takes a need (for testes to be kept cool), adds to it his own teleology (the testes were somehow forced to move), and so creates a solution out of nothing.

One should not miss the vague nature of his language. "Likely," "We don't yet know," "a clue," "possible," and "perhaps" all account for something that Fodor and Piattelli-Palmarini pointed out. Arguing from the model allows one to be vague because, very often, there is not enough data available to fill in the model. At this point we might say that the model is underdetermined. This is how Fodor and Piattelli-Palmarini (FPP) describe the situation

Still, this book is mostly a work of criticism; it is mostly about what we think is wrong with Darwinism. Near the end we'll make some gestures towards where we believe a viable alternative might lie; but they will be pretty vague. In fact, we don't know very well how evolution works. Nor did Darwin, and nor (as far as we can tell) does anybody else. 'Further research is required', as the saying goes. It may well be that centuries of further research are required.[98]

Their (FPP) "why" is that natural selection is not a workable principle. The scope of their effort is to show how its parallel was rejected in B. F. Skinner's work and, if it is to be rejected in one then why not the other? Why not even some stronger criticism?

So, if it is true that Skinner's theory and Darwin's are variations on the same theme, why aren't the objections that are routinely raised against the former likewise raised against the latter? If nobody believes Skinner any more, why does everybody still believe Darwin?[99]

The question raised by Fodor and Piattelli-Palmarini is a call for consistency. It is not a call to reject data. It is a call to reject a broken model. If the model is broken is it not broken? It seems silly to state it that way, but that's what their argument boils down to.

On the Structure of the Arguments

This examination of the arguments of the scientist includes an overview of the epistemology, a summary of the argumentation methods, some philosophical concerns, an assessment of model exclusivity, and some concerns about the particular content each model.

On Contextualist Arguments

Before digging into the philosophy statements about contextualism as applied here, let's begin with a summary of the problem. What it boils down to is a sophisticated case of question begging. It is tempting, and sometimes happens without one being aware, that the assumption of an argument is added to the conclusions. In this case naturalism is held to as the reason for naturalistic evolution and naturalistic evolution is taken as proof or evidence of naturalism. Even more briefly, evolution accounts for naturalism and naturalism accounts for evolution.

This first section simply sets up some ground rules for justifiability. Michael Williams provides this discussion[100] regarding possible, and realistic, criticisms for claims that may be inappropriately justified:

It is tempting to think of the presuppositions that are held in place by the direction of inquiry and other contextual factors as constituting a kind of 'framework' within which justificational questions can be raised and answered. I think that this way of talking is best avoided. It is almost bound to lead to contextualism's being confused with the relativist thesis that all justification is 'framework-relative', in a way that places frameworks themselves beyond justification or criticism. . For example, Fogelin represents the contextualist as claiming:

(FR) S is justified in believing that p if p is justified within the framework in which S is operating.

However, as Fogelin points out, there are at least three ways of resisting the move from a claim to be justified 'within a framework' to a claim to be justified simpliciter:

(a) I may reject S's justificatory framework. (S may be using astrological tables.)

(b) I may accept S's justificatory framework, but think S has not used it correctly.

(c) I may grant that S has been epistemically responsible, but think his grounds have been defeated.

Fogelin rejects contextualism because he takes it to exclude critical reactions like these. This is a mistake: a contextualist can and should accept the potential legitimacy of all three critical moves.

A contextualist view of knowledge and justification does not commit one to holding that a reference to context is part of the *content* of a knowledge-claim. We must recall the *sotto voce* proviso. A knowledge-claim commits one to holding that all significant potential defeaters — possibilities which, if realized, would make one's belief either false or inadequately grounded—have been eliminated: the contextual element comes into fix what defeaters should be counted significant. But presuppositions as to what is significant are themselves open to criticism, which can be informationally or economically triggered.

What I take from Mr. Williams, especially noting the concluding sentence, is that a framework (a presupposition or a system of assumptions which are used as a framework) is not in and of itself adequate to determine the justifiability of claims.

It appears to be the case that the naturalist evolutionist will frame evidential arguments using a faulty contextualist approach as justification is sought for naturalism. The end is that the assumptions (such as common descent) are treated as both justifiable and justified. To do this, though, requires (according to Williams) that all defeaters have been addressed. But has this been accomplished? I do not know believe that it is necessary to offer a criticism on this basis from an atomistic approach (critiquing each bit of evidence). The alternative approach is to show the failings of the model structures themselves, and that leaves the evidence open for reframing. The question of *how* comes first, and answering that will lead to inquiring whether or not the naturalist has accomplished his goal.

Now return to the first two quotes of this effort, from Coyne and Prothero. Coyne's broader assertion is that[101]

> Again one must ask: If animals were specially created, why would the creator produce on different continents fundamentally different animals that nevertheless look and act so much alike? It is not that marsupials are inherent superior to placentals in Australia, because introduced placental mammals have done very well there. ...

> No creationist, whether of the Noah's Ark variety or otherwise, has offered a credible explanation for why different types of animals have similar forms in different places. All they can do is invoke the inscrutable whims of the creator. But evolution *does* explain the pattern by invoking a well-known process called *convergent evolution*.

Behind these remarks there is that strong dependence upon evidence, in this case to prove common descent. It is so strong, in fact, that for Coyne the evidence is sufficient in and of itself to prove the natural processes, given of course that the correct assumptions are put in place to deal with the evidence. In this case he employs the assumption of convergent evolution to provide the power to leap from the evidence to his naturalism.

This assertion of evidence is accompanied by the assertion of a defeater for creation. He speaks of the apparent lack of a creation explanation in the various scenarios for the diverse locations of animals having similar features. This would also seem to assume that if creation did not occur in one place then it did not occur in any place, since it is claimed to be a universal event.

Consistent with this, Donald Prothero[102] asserts:

> By now the evidence for evolution is overwhelming, so the burden of proof on the antievolutionist is much larger: they must show creationism is right by overwhelming evidence, not just simply point out a few inconsistencies or problems with evolutionary theory.

Here Dr. Prothero takes an even stronger evidential position. But not only does he appeal to the content of the evidence, he also appeals to the evidential method. His challenge is that we must prove creation by "overwhelming evidence" and by no other approach. But oddly enough his sense of knowledge is hardly consistent.

Dr. Prothero has thrown down the gauntlet by demanding that the opposition employs, if possible, a univocal evidential approach while he retreats to the safety of his presuppositional castle where the "naturalistic explanation" is sufficient. His contextualist approach assumes that the evidence speaks for itself and that, if one only looks at the evidence, then one cannot help but come to the same conclusion that the naturalist arrives at.

The defeater that he raised for the creationist is one of evidence. Though this was not explicit in Dr. Coyne's statement, it was none-the-less implicit. The challenge is to come up with new evidence to prove something different– to justify that new claim of creation.

On Human Capacities

The way that evidence is managed comes with the assumption that one's faculties are suited to justify these belief systems. Alvin Plantinga argues that a purely naturalistic epistemology differs enough from a Christian epistemology that the trust in human faculties that the naturalist demands, this cannot be attained.[103] As an alternative he promotes a practical approach to warranted belief by promoting an epistemology, still naturalistic, that acknowledges and employs the historical context of theism that surrounds Western thought:

> So the view I propose is a radical naturalism: striking the naturalistic pose is all the rage these days, and it's a great pleasure to be able to join in the fun. The view I urge is indeed best thought of as an example of naturalistic epistemology; here I follow Quine (if only at some distance). Naturalistic epistemology, however, is ill named. In the first place, it is quite compatible with, for example, supernaturalistic theism; indeed, the most plausible way to think of warrant, from a theistic perspective, is in terms of naturalistic epistemology. And second ... naturalism in epistemology flourishes best in the context of a theistic view of human beings: naturalism in epistemology requires supernaturalism in anthropology.

So what difference does this make? Plantinga's conclusion is that a purely naturalistic worldview limits the perception of reliability that the scientist might attain. To resolve this problem he introduces a "radical" natural epistemology that operates within the realm of theistic epistemology.

Of course the level of trust varies among scientists. Some, as we saw with Prothero, will first maintain that the evidence is more than adequate to justify the evolutionist's position. But the next leap – from the evidence for speciation as evidence for naturalism as a necessary conclusion – may not be quite so strong.

Analogical and Univocal Arguments

Analogical-presuppositional discussions present their arguments as flowing from the presupposition out to the evidence and framing the evidence according to the presupposition and its immediate assumptions. That is, they will justify the evidence (both theory structure and interpretation) from the presupposition. In the case of evolutionary theory, this is generally from the position of naturalism and dovetails the contextualism issue. Here the theorist will begin with the evidence and attempt to conclude not only naturalistic origins (common descent) but even the presupposition naturalism itself.[104]

Of course the use analogy is a legitimate method of argumentation. It allows word pictures to be presented so that the listener gets a proper picture of the point being made. Analogies may also establish relationships and correlations between seemingly disparate entities so that similarities might become clear. The use of analogy is not of concern but rather the method in which it is used. Is it being used in a consistent fashion and do the analogies represent fairly the subject being clarified? The following are some examples of this approach.

Stephen Jay Gould

Of interest is Gould's discussion of analogous vs. homologous development. Exploring the specific matter of Drosophila segmentation, he approaches the matter as so

> (3) Does segmentation occur in developing vertebrae, and could such a process be homologous with the conversion of embryonic insect parasegments to adult segments? As De Robertis (1997) reminds us, anatomical data known for more than a century indicate that a subset of cells in each somite (called the sclerotome) forms a vertebra. But each adult vertebra arises by "resegmentation" as the posterior half of one sclerotome fuses to the anterior half of the next sclerotome along the A-P axis. "The end result is a phase shift of the vertebra with respect to the muscle, so that the segmental muscles can span, and move, adjoining vertebrae" (De Robertis, 1997).

> These anatomical data, never satisfactorily verified, have now been confirmed by cell lineage studies in birds. De Robertis argues that such vertebral resegmentation may be homologous, and not merely analogous, with the similar construction of insect segments from conjoined halves of adjacent parasegments. De Robertis concludes (1997): "It seems improbable that such a complicated way of making individual metameres would have arisen independently twice in evolution." [105]:

To this point one must respond regarding Gould's other related statements. For instance, the principles of both parallel and convergent[106] evolution hold a strong position in his model. It seems odd that equally complex structures would develop in one case only once and in other cases in parallel. Consistent with this, his tentative conclusion is that homology is limited[107].

He thus chooses homologous development[108] with respect to eyes and vision, and this raises some serious questions about the processes, or at least about potential models to cover the subject. His solution to the development of vision in some creatures requires a leap from a disconnected light sensor to a meaningfully-connected light sensor. He asks, does any one eye equal any other eye? Not according to his model. He says:

> This case has persisted as a classic, ever since the formulation of convergence as a concept, because the two eyes look so much alike, and work so similarly, despite their separate origins form different tissues: the vertebrate eye as an evagination of the brain, and the cephalopod eye by invagination of the epidermis. The squid eye forms from a monolayer of epidermis that becomes thickened, multilayered, and internalized in the dorsal side of the head lobe. [109]

In one sense this is an argument from a practical standpoint. He is asking what works best and he seeks the answer that makes the most sense. Why should evolution, he suggests, do things the hard way? Why should it not choose the simplest path for each organism? Could it not be that there are multiple paths to the same end, he might ask? That is, in fact, what he is asking.

The analogy here breaks down. What analogy? It's the analogy of the different creatures and their assumed differing evolutionary development processes. Though all may have eyes not all seemingly developed this sense of vision through the same means and mechanisms.

Gould's analogy and argument proceed from the principles of evolutionary theory. But contrast that to his description of drosophila segmentation[110] and the development of vision. He does not begin with the presupposition. He begins with the evidence, or in this case the model that he treats as evidence (that is, his understanding of the origins of the eyes), and then proceeds to use the evidence to support the assumption of convergence. In these instances he reverses his methodology.

This may appear an arbitrary move. Taken in isolation from the evidence, it is. But the model drives this. It is a supportive structure that the model demands in order to function properly. The problem, then, is that this sub-model analogy is treated as the equivalent of evidence. The analogy becomes one of the facts, and that's the problem.

This statement is made in the context of a criticism of ID and its apparent absoluteness that some complexities *cannot* have come about by means other than special creation. To that he provides an alternative:

> We need only show that such a development, involving processes and constituents not unlike those we already know and can agree upon, is "feasible." And by "feasible," they mean that there must be evolutionary precursors of each new trait, and that evolution of that trait does not violate the Darwinian requirement that each step in building an adaptation benefits its possessor.[111]

Then he proceeds to provide an example of how these things might occur:

> Although advocates of ID claim a supernatural hand behind these pathways, dogged scientific research is beginning to give plausible (and testable) scenarios for how they could have evolved. Take the blood-clotting pathway of vertebrates. This involves a sequence of events that begins when one protein sticks to another in the vicinity of an open wound. That sets off a complicated cascade reaction, sixteen steps long, each involving an interaction between a different pair of proteins and culminating in the formation of the clot itself, although more than twenty proteins are involved. How could this possibly have evolved?

> We don't yet know for sure, but we have evidence that the system could have been built up in an adaptive way from simpler precursors. Many of the blood-clotting proteins are made by related genes that arose by duplication, a form of mutation in which an ancestral gene, and later its descendants, becomes duplicated in full along a strand of DNA because of a mistake during cell division. Once they arise, such duplicated genes can then evolve along separate pathways so that they eventually perform separate functions, as they now do in blood clotting.[112]

This leaves Dr. Coyne with a little problem. Is the directionality of evolution now dependent upon presumed accidents, genetic infections, and other unknowns for its sense of certainty? The level of dependency here is almost as deep as the protein action levels of blood clotting. But more important is the argumentation method. Dr. Coyne is actually arguing, not from the evidence, but from the model that frames the evidence. His analogy is the model and its relationship to the evidence. If a trait occurred in one instance, why could it not, he suggests, have occurred by the same mechanisms in another instance?

To be fair, his argument continues. He looks for a parallel that might meet the demand of the model.

> Although an intelligent designer could invent a suitable protein, evolution doesn't work that way. There must have been an ancestral protein from which fibrinogen evolved.

> Russell Doolittle at the University of California predicted that we would find such a protein, and, sure enough, in 1990 he and his colleague Xun Xu discovered it in the sea cucumber, an invertebrate sometimes used in Chinese cooking. Sea cucumbers branched off from the vertebrate lineage at least 500 million years ago, yet they have a protein that, while clearly related to blood-clotting proteins of vertebrates, is not used to clot blood. This means that the common ancestor of sea cucumbers and vertebrates had a gene that was later co-opted in vertebrates for a new function, precisely as evolution predicts.

This is one point where the model begins to break down, though ironically it is also a place where the naturalistic evolutionist seems firmly planted. Once again he remains firmly analogical and argued from his assumptions and not from the evidence.

A myriad of assumptions are made about ancestry, the function of the genes, the permanence of the inherited genes, the origin of the inherited genes, and the ability of the genes to initiate a function. But how does it break down? His argument is that the gene, while existing, has the same origination point and serves, or served, the same function at some time in the past. There is no pursuit of any parallel or convergent theory option – that perhaps the origin might have differed from his model. Why? *Because the model mattered most* and further analysis of the issue becomes undesirable.[113]

Dr. Coyne's introduction provides an example of the way model assumptions might feed into themselves.

> Actually, the nested arrangement of life was recognized

long before Darwin. Starting with the Swedish botanist Carl Linnaeus in 1635, biologists began classifying animals and plants, discovering that they consistently fell into what was called a "natural" classification. Strikingly, different biologists came up with nearly identical groupings. This means that these groupings are not subjective artifacts of a human need to classify, but tell us something real and fundamental about nature. But nobody knew what that something was until Darwin came along and showed that the nested arrangement of life is precisely what evolution predicts. Creatures with recent common ancestors share many traits, while those whose common ancestors lay in the distant past are more dissimilar. The "natural" classification is itself strong evidence for evolution.

Why? Because we don't see such a nested arrangement if we're trying to arrange objects that haven't arisen by an evolutionary process of splitting and descent.[114]

For Dr. Coyne it seems that observations themselves are evidential. It appears that he is confusing the presuppositions and assumptions with the examined evidence.

In that light his definition of science is of interest.

This brings us to the second point. For a theory to be considered scientific, it must be *testable* and make *verifiable predictions*. That is, we must be able to make observations about the real world that either support it or disprove it. Atomic theory was initially speculative, but gained more and more credibility as data from chemistry piled up supporting the existence of atoms. Although we couldn't actually *see* atoms until scanning-probe microscopy was invented in 1981 (and under the microscope they do look like the little balls we envision), scientists were already convinced long before that atoms were real. Similarly, a good theory makes predictions about what we should find if we look more closely at nature. And of those predictions are met, it gives us more confidence that the theory is true.[115]

By this standard a theory does not become scientific until it meets the two stated criteria – it must be testable and verifiable. The problem is that his standard, just one page earlier, is that evolution is a scientific theory. And this theory is built upon the assumption that "if speciation is true, for instance, then common ancestry must also be true." But that *must* is an unverifiable assertion. So, either his definition of evolution is lacking or evolution is unscientific. (Of course it remains with me to show that the assumption is unverifiable.) So in this he is again being analogical rather than evidential in his argument.

Dr. Coyne makes another set of statements with the same self-feeding form as was first noted.

> Evolutionary theory, then, makes predictions that are bold and clear.

> As we'll see, all the evidence – both old and new – leads ineluctably to the conclusion that evolution is true. [116]

This statement takes the reverse approach. Here he says that the evidence comes to only one inescapable conclusion. So he reverses his other statements, a statement that sits juxtaposed to his other statements.

In an article that he wrote for *The Nation*, Dr. Coyne reinforces this direction. Not being evidential, but rather analogical, he talks about evidence but does not appeal to his evidence: He appeals to his presupposition of an empty cosmos composed only of the natural "blind and mindless process of natural selection" while concluding with his appeal to "multifarious evidence."

> One answer is religion. Unlike germ theory, the idea of evolution strikes at the heart of human ego, suggesting that we were not the special object of God's attention but were made by the same blind and mindless process of natural selection that also built ferns, fish and rabbits. Another answer is ignorance: most Americans are simply unaware of the multifarious evidence that makes evolution more than "just a theory," and don't even realize that a *scientific* theory is far more than idle speculation. [117]

Dr. Coyne's analogy of the universe to "blind and mindless" processes remains an unverifiable, idle speculation.

A Matter of Character

Fair criticisms, while not always pleasant, are beneficial to all. Unfair criticisms, criticisms that misrepresent positions or miss the point entirely are ones which benefit no one. For instance, in the 19th century the abuse that Samuel Wilberforce heaped on Thomas Huxley did no good for the advance of Christianity against naturalism. That example serves many today as an example of how Christians appear "anti-science" in their position.

A more recent example is Jerry Coyne entirely missing the point being made by Nancy Pearcey. After quoting her concern about naturalism and the place of ethics he says that

> Pearcey argues (and many American creationists agree) that all the perceived evils of evolution come from two worldviews that are part of evolution: naturalism and materialism. Naturalism is the view that the only way to understand the universe is through the scientific method. Materialism is the idea that the only reality is the physical matter of the universe, and that everything else, including thoughts, will, and emotions, comes from physical laws acting on that matter. The message of evolution, and all of science, is one of naturalistic materialism. Darwinism tells us that, like all species, human beings arose from the working of blind, purposeless forces over eons of time. ...

> But Pearcey's notion that these lessons of evolution will inevitably spill over into the study of ethics, history, and "family life" is unnecessarily alarmist. How can you derive meaning, purpose, or ethics from evolution? You can't. ...

> But evolution can still shed some light on whether our morality is constrained by our genetics. [118]

His discussion goes further and does address the concern about ethics more specifically. More on that in a bit. Here Dr. Coyne did several things. They are, not necessarily in order, (1) misrepresent the history of scientific inquiry, (2) minimizes Pearcey's concern, (3) minimizes Pearcey personally, and (4) trivializes the actual issue. This approach is not "science" in any sense. Nor is it an effective apologetic. It amounts to avoiding the issues by diminishing one's opponent.

One might reasonably infer from Dr. Coyne's expressions that any argument which evolution proposes regarding ethics through evolutionary means is evidentially and empirically false.

It seems that we live in a world where the old Christian morality is assumed. It might be argued as a consequence that much of the ethic proposed by naturalistic evolution is tainted by the Judeo-Christian ethic of our world. This ethic was so entrenched in European and American society that philosophers have assumed this accepted foundation for ethics would produce the relatively peaceful society we know today. They're taking credit for what they did not, and could not, create.

They do not often recognize that this is not one world and that the many variant ethical systems produce different results. Our world is filled with many worldviews, each holding to a different standard for their ethic. The standards many scientists hold are, just as was Darwin's theory, founded in Christian theology, rather than Reason apart from Christianity. It is difficult to avoid the ethics question entirely. Darwin had more than one doubt. So did his contemporaries.

> When Spencer, also with second thoughts, began to assail English militarism and to predict the degeneracy of the nation through the rise of a new serfdom, he was scouted as a pacifist, a dotard, and a dissenter. Forty years before, on the appearance of the Origin of Species, a Manchester journalist had shown that according to the author's scheme "might is right and therefore Napoleon is right and every cheating tradesman is also right." Darwin had not understood. But in the late seventies he had grown alarmed. "I am beginning to despair of ever making the majority understand my notions." This from the man who had "conquered the opinion of the world." No. The gladiatorial conception of the struggle for existence was there to stay, and as Geoffrey West rightly says, harking back to old Sedgwick and his complaint about neglecting the moral and metaphysical part of man: "He leapt ... right to the heart of the matter in a prophetic passage whose insight should be more apparent now, when the increasing brutalization and degradation of humanity are no more to be denied than detached from conception of evolution and natural selection..."[119]

The materialism of Darwin created an internal doubt. He did not doubt the subject of his inquiry but about the character of his theory. His was the age of materialism, of Hume's empiricism, of the materialism of Marx, and mechanization of the Industrial Revolution. Materialism had shown itself to be bereft of those things that had traditionally defined what we thought of as "human." The peace of the Christian world that Darwin grew up in, and practiced in, had now lost its foundation.

Today's radical scientists, the "new atheists," continue to pursue this doubt. "The Moral Landscape" is such an effort. Here Sam Harris attempts in vain to get an "ought" (a moral imperative) from and "is" (a materialistic world). But all he can do is appropriate for the reader an empty hope as he tries to reclaim the moral foundations that were destroyed by materialism.

The Engines of Evolution

Naturalistic evolution has frequently been challenged about the legitimacy of its structure. Barzun, Butler, Polanyi, and Nagel have raised important and real concerns that are yet to be answered adequately. The question of materialism is a foundational concern and is seen by many as necessary for maintaining Darwinism.

It is assumed that materialism is capable of producing life. It is the *how* of this that comes next. The question is: *How does materialism function to accomplish these things?* From this we get the subsequent questions: *How are genes modified? How are traits developed? How do species develop?* And so forth.

Darwinism and neo-Darwinism make their appeal to two concerns: Behavior and genetics. Which one gets priority depends upon the model in use. It is at this point we diverge from the standard approach. One will usually hear discussions of "the scientific model" and assume that there is but one. As we have already seen there are at least two – Darwinism (adaptation) and neo-Darwinism (genetics + adaptation).

Darwin recognized that traits were developed as a response to the environment and the entity's adaptation to that environment. His was a needs-based approach. But it failed to answer the question of how the traits changed in order to produce the needed feature.

Then along came the neo-Darwinists. Huxley, Mayr, Haldane, Richenbach, et al, found that perhaps the answer to these questions comes about differently. Perhaps it is the genes that create the feature and this feature creates the strength to adapt. For some this might come about slowly (based on evidence found in the fossil record, aka phyletic gradualism) while for others this might come about rather quickly (based on the newer discoveries on how genes might behave, aka punctuated equilibrium). There are respected scientists who are aligned with both camps.

Next we encounter a more skeptical approach. What if sometimes, it is proposed, that adaptation is the primary force at work and at other times the change in genetics is the primary force (Fodor/Piattelli-Palmarini)? It's an interesting question, especially as it seems to allow the strengths of each major camp to present its best case, thus seemingly cementing in place a model that has the capacity to answer all of the questions. By taking this approach it would appear that the disagreements have been resolved. Yet other questions remain.

Outside of Darwinism there are the non-Darwinists. Some like Thomas Nagel have suggested that, while naturalistic evolution is true, Darwinism's materialism is incapable of answering certain questions. (Fodor & Piattelli-Palmarini raise the same question.) For him the model does is not capable of handling all the data and as such needs to be changed or replaced. In that light we may add this as the fifth category. Yet because it is at this point much of what we read is a response to Darwinism, rather than a full model meant to take its place, it still falls under the camp of Darwinism (as it panders to it), or is at least connected to it as it shares some of Darwinism's constructs.

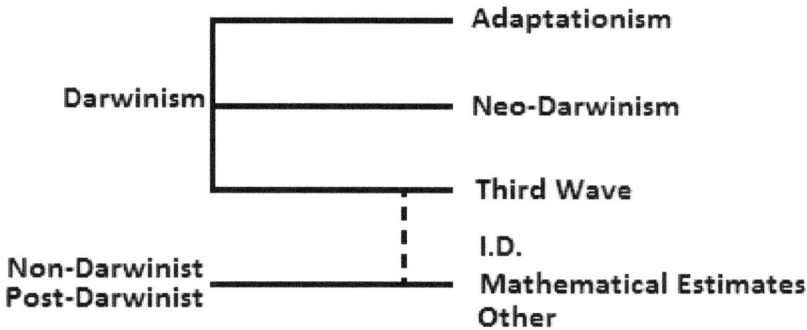

```
                         ┌──────────── Adaptationism
              Darwinism ─┤──────────── Neo-Darwinism
                         └──────────┐
                                    ┊   Third Wave
                                    ┊
       Non-Darwinist                ┊   I.D.
       Post-Darwinist ──────────────┴── Mathematical Estimates
                                        Other
```

Identifying Model Differences

The model differences may be seen in how certain information is treated. This is where the differences standout. One model claims behavior is enough while the other observes that genetics followed by adaptation is what drives the system. Though they agree at the outer level regarding common descent, past that point they oppose each other regarding how evolution works.

Within the various strains of neo-Darwinism there are also conflicts. Haeckel's recapitulation theory, the idea that previous stages of evolutionary development may be observed in the development process of the embryo, serves as an example. Jerry Coyne says:

> The "recapitulation" of an evolutionary sequence is seen in the developmental sequence of other organs – our kidneys, for example. During development, the human embryo actually forms three different types of kidneys, one after the other, with the first two discarded before our final kidney appears. And those transitory embryonic kidneys are similar to those we find in species that evolved before us in the fossil record – jawless fish and reptiles, respectively. What does this mean?
>
> You could answer this question superficially as follows: each vertebrate undergoes development in a series of stages, and the sequence of those happens to follow the evolutionary sequence of its ancestors. So, for example, a lizard begins development resembling an embryonic fish, then somewhat later an embryonic amphibian, and finally an embryonic reptile. Mammals go through the same sequence, but add on the fin stage of an embryonic mammal.
>
> This answer is correct but only raises deeper issues. Why does development often occur in this way? Why doesn't natural selection eliminate the "fish embryo" stage of human development, since a combination of a tail, fishlike gill arches, and a fishlike circulatory system doesn't seem necessary for a human embryo? Why don't we simply begin development as tiny humans – as some seventeenth-century biologists thought we did – and just get larger and larger until we're born? Why all the transformation and rearrangement?
>
> The probable answer is – and it's a good one – involves recognizing that as one species evolves into another, the descendant inherits the developmental program of its ancestor: that is, all the genes that form ancestral structures.
> 120

Coyne explained recapitulation by way of inherited genetic content. What he really did was to substitute one language for another. Gone is the simplicity of observed stages and the appearance of something that looks like gills and in its place is the simple language "one species evolves into another" and assumes that evolution is the source.

Alternatively Mark Isaak's *The Counter-Creationism Handbook* raises an important question as he attacks a particular criticism of recapitulation:

> The pharyngeal pouches that appear in embryos technically are not gill slits, but that is irrelevant. The reason they are evidence for evolution is that the same structure, whatever you call it, appears in all vertebrate embryos. Agassiz (not a Darwinist himself) said, "The higher Vertebrates, including man himself, breathe through gill-like organs in the early part of their life. These gills appear and give place to lungs in a later phase of their existence" (Agassiz 1974).

> Darwinian evolution predicts, among other things, similar (not identical) structures in related organisms. That pharyngeal pouches in humans are similar to pharyngeal pouches (or whatever you call them) in fish is one piece of evidence that humans and fish share a common ancestor. [121]

The question that arises has to do with scientific methodology. Do we merely observe external characteristics and presume that similarity equals history? (That's a weak analogy.) If these are not gills, as Isaak acknowledges, then what are they? Are they not merely a structure that is similar in *appearance* yet differing in *function*? Isaak acknowledges that but seems unwilling to accept it as a contradiction.

Gould's criticism of this biogentic law raises a serious concern:

> The phylogeny of a lineage unfolds through thousands of steps in geological immensity; new sages cannot be added indefinitely to the unaltered ends of previous ontogenies, lest growth to adulthood take untold years to reach completion. Some process – some law of heredity – must produce a general speeding-up of development, so that ancestral ontogenies can unfold more rapidly, leaving time at the end for addition of novel features. [122]

That leaves the scientist who accepts even the fundamental concept of Haeckel's theory with a serious problem. Not only do they generally contradict their understanding that some sort of factual proof comes from their science, but also Gould's issue of teleology makes the theory, as it stands, unworkable. That is, it is no longer "random" and one of its first principles appears to be *de facto* falsified. Now Gould, to be fair, made clear that he thought there might be some qualified future use for a modified version of this idea[123]. Yet what remains is a functional teleology. Randomness in any sense does not fit well with teleology.

In this case Gould's approach is better able to handle the biogenic question. He recognized the issue of teleology and dealt with it accordingly. But is it fair to say that his model structure produces a better analysis of the data than does Coyne's? The better question might be why Coyne would stake a claim using such an outmoded theory unless he believed that such a theory were suitable support for the model.

On the issue Ernst Mayr says:

> When evolution was accepted, a new definition offered by Haeckel (1866), "Ontogeny is the recapitulation of phylogeny," obviously went too far, because at no stage of its development does a mammalian embryo look like an adult fish. Yet, in certain features, as in the gill pouches, the mammalian embryo does indeed recapitulate the ancestral condition. And such cases of recapitulation are by no means rare. The larvae of barnacles are very similar to those of other crustaceans, and embryonic structures are found in thousands of cases to be indicative of their ancestry, but these same structures are absent in the adult life forms. [124]

Mayr makes the same error as did Lamarck: He argued from resemblance. He supports the position from within the narrowness of a set of similar creatures instead of verifying that the so-called *gill slit* that he calls a *gill pouch* is actually a recapitulation of something from the past. The demand for verification of a scientific position sometimes takes a back seat to the systematic.

In this case the punctuated equilibrium (PE) model Gould is able to produce a better answer to the question than is either Coyne's or Mayr's approach. They sit in conflict, with one being in clear error and the other being potentially correct and only partially in error.

But Gould has not escaped the teleological issue. He is willing to accept, without question, a certain directionality though rejecting teleology. After beginning a discussion by interacting with how Job might be interpreted, he delves into scientific interpretation:

> I chose this unconventional mode of beginning a scientific discussion with Biblical exegesis because this chapter (and a central theme in the logic of this entire book) rests upon a

particular definition and construction of the concept of constraint – a meaning easily defended both terminologically and factually, but often so buried in a confusing and contentious literature that the centrality of the argument becomes lost in collegial frustration. [125]

What Gould sought was an explanation for how things work that separates the mechanics from any external influence. As such his use of *constraint* is, as he says, a positive one. He sees as the whole of the process as "promoting change in particular directions." [126] But this is directionality is a big piece of what is meant by teleology. Any sense of promotion or any sense where some external factors determine the path – these lead to an end. It may not be a theological end and it may not have a measurable end point. Yet each predictable step in a direction is a predictable end or purpose.

These things are not necessary in the discussion. What is important is that teleology, at some level, is implicit in directionality. While the various models conflict with respect to recapitulation and with respect to claims about directionality they also conflict with regard to how they function. In this case the PE model appears to contradict itself in order to handle the evidence of the moment. It would also appear that, while nobody in the scientific world wants teleology, the principle may be inescapable.

The differences in how each model functions are also part of the claims each makes. It need not be inferred by outside observers. Gould made these differences explicit. One issue is to find what defines a change in species. For Gould it is the group; for Dawkins, it is the individual. Gould says that

> Darwin's real theory of natural selection is a theory of ultimate individualism. Darwin's mechanism works through the differential reproductive success of individuals who, by fortuitous possession of features rendering them more successful in changing local environments, leave more surviving offspring. Benefits accrue thereby to species. [127]

Gould's explanation bears more similarity to Mayr's proposal. In contrast to this Dawkins says that

> Genes are competing directly with their alleles for survival, since their alleles in the gene pool are rivals for their slot on the chromosomes of future generations. Any

gene that behaves in such a way as to increase its own survival chances in the gene pool at the expense of its alleles will, by definition, tautologously, tend to survive. The gene is the basic unit of selfishness.[128]

Gould and Dawkins do not disagree on speciation. They only disagree on the *how it works* question. Dawkins' principle seems closer to the gradualism of Darwin but Gould's is faster in order to answer some of the genetic and paleontological concerns.

At this point we have seen a list of differences. Some might be tempted argue that the process has evolved to become what it is today. That would be true if there were consensus. But one can find scientists who agree more with Gould, or more with Mayr, or more with Dawkins, or more with Darwin. Why do they disagree? That was the point made, as noted earlier, by Fodor and Piattelli-Palmarini – nobody knows how it works.

That challenge goes back to questions surrounding beginnings, of how life began to evolve. These are about the origin of life and the behavior of early life. *Which came first, life or DNA? Does DNA create the situation where life might advance or does life create the environment where DNA might contribute?* It is a real chicken-egg debate, as well as a serious teleological concern. Hubert Yockey says it this way as he interacts with Commoner on the subject. First he quotes Commoner and then clarifies the point that

> Because of their commitment to an obsolete theory, molecular biologists operate under the assumption that DNA is the secret of life, whereas careful observation of the hierarchy of living processes strongly suggests that it is the other way around: DNA did not create life: life created DNA.

> [Now he really steps on his argument:]

> When life was first formed on Earth, proteins must have appeared before DNA because, unlike DNA, proteins have the catalytic ability to generate the chemical energy needed to assemble small ambient molecules into large ones such as DNA

> This quotation shows that Commoner does not understand the genetic information system and the Central Dogma. It is mathematically *impossible*, not just *unlikely*, for information to be transferred from the protein alphabet to the mRNA

alphabet. That is because no codes exist to transfer information from the twenty-letter protein alphabet to the sixty-four-letter alphabet of mRNA. [129] (emphasis in original)

In the end the issue of teleology goes unresolved. With one you get directionality and with the other teleology. It's a sticky situation and neither position is to be envied for the challenges it faces.

The consequence is not just that some molecular biologists do not understand the full scope of the evolutionary system. Dr. Yockey thus has some important information to contribute to the greater discussion. He has just opened the door to the most difficult question of all, that of the origin of life. The questions revolving around DNA and RNA remain challenging ones. There appears not to be a current model that might answer the challenge without serious structural issues, especially of teleology.

The Question

At this point we question which model is truest – which gives the best answer to the question of accounting for biological history? Which handles the data best, and comes closest to representing reality most precisely? For this matter the argument the model must be first a *valid* argument – it must be structured properly. Then it must be a *sound* argument — its premises are true. And finally it must have a positive truth value – something *demonstrable*.[130] Even with all of these things in order an argument may still be in error, for these are matters of structure and not of the truth-value of the argument itself.

In the case of the various evolutionary models it seems that the first criterion is met. But after that, with the level of vague language that persists, it is difficult to call the arguments sound. Of course some demonstrable result (fruit) has been brought. But the fruit does not validate the vague language on which it was built.

The issue of vague language also speaks to what Fodor and Piattelli-Palmarini indicated, as did Polanyi several decades years prior. Vague language does not make for good theory. Vague language might fairly be branded as a *guess*.

IV. A Structured Defense

I trust that by this point you have seen some of the issues. Not everything is about evolution per se but about the structure of the arguments. The structure is driven by a naturalistic worldview with its demands. While much of this apologetic is done in the context of evolution and origins it is more concerned about the structures than about the content itself.

Criticisms of evolutionary theory, I believe, are not effectively accomplished by looking at errors in how some of the detail is handled. That subject was dealt with previously. It seems to me that taking that approach gets us nowhere and advances neither God's kingdom nor good science. Instead, with one's eyes on the goal of doing the best science possible, we may then present those criticisms of naturalistic evolution theory structures which are constructive to good science and good theology.

The Character of this Apologetic

In the mid twentieth century two philosophers, Paul Feyerabend at Berkeley and Cornelius VanTil at Princeton (and subsequently Westminster), confronted the assumed correctness of the movement called "logical empiricism," also known as "logical positivism." The ideas that they expressed were and are not popular in many circles but they maintain some significance in both science and theology.

Both of these thinkers took a functionally presuppositional approach. That is, they looked at the character of the person making the formula and drawing the conclusions and in doing so would critique the character of a theory. VanTil proposed (among other things, of course) that it is the fallenness of humanity that taints knowledge. Nobody can account for knowing anything apart from the revelation given by God. Of course we don't get mathematics and such from God. The Bible isn't a math book. But it is how we account for knowledge that's important here. Our ability to know is a part of general revelation.

We generally attribute to general revelation things like the grandeur of mountains or the simple presence of joy in one's life. It is easy to take Paul's use of the term "nature" in Romans chapter one and think only of the physical world. I would suggest that the idea should be expanded to include all of the abstractions such as mind, logic, and creativity as elements of creation and of the nature that was created. For the Christian all is part of God's creation, God's domain. It is on the level of general revelation where we share observation and participation with the rest of the world. But the world does not account for the breadth of nature in the same way.

The current naturalistic response to this proposition might be: *Naturalistic and materialistic processes account for the mind's perceptions. What the mind perceives requires no deity or spirit. It requires only the capacity to perceive. We all agree and account for "red" simply by consensus. The remainder is merely conjecture.*[131] While at first compelling, a statement of this type ends as question begging. It is assumed that the mind was the result of natural processes and then the mind is used to assess these apparent natural processes. But we have to understand how the mind might account for itself. How might the mind find its own beginning? The cycle is endless.[132]

To account for something is not simply to assert "God" and leave it at that, though this accusation is rather common. It is, rather, to find the starting point for knowledge. It is to pursue the beginning, just as natural science asserts to do.

In contrast to this the current philosophical trend in apologetics employs the methods of the Rationalist, often employing the methods of Aquinas and Aristotle, to identify Christianity as superior. While the details of this approach can be valuable the framework leaves something to be desired.

It works like this: It is assumed that Reason is sufficient to show that one ought to choose Christ over naturalism, Islam, Hinduism, or any of the other competing systems. Christianity, being the most rational and the most reasonable, is the inferred best conclusion. The problem here is fundamental: What happens if someone, based on the arguments presented, can find a better answer, according to reason (using the evidence at hand), than the Christian worldview or theological system? If Reason is assumed to be correct and consistent within the human experience, then it appears that Reason trumps revelation as the source for truth and that accounting for Reason is not necessary. It is thus possible for the Rationalist-type of Christian apologist to find a reason to doubt his faith through his faculties. (I've not known of anyone to do this – to reject Christ on account of a philosophical structure. But yet the structure remains and thus the concern.)

This presuppositional approach might better be termed a "revelation-based" approach, as revelation is the first source for truth. K. Scott Oliphint, a current presuppositional apologist, has taken a slightly different approach and renamed it to "covenantal apologetics"[133] on account of revelation's relational and covenantal interaction with humanity.

On the other hand, Feyerabend's approach was most certainly not based in revelation. His approach seems typical of Berkeley in that era and his approach is sometimes called "anarchist." What he did was question how objective the mind could be. One might easily assume that a good deal of our knowledge is obvious. Part of this is wrapped around what we can touch and test ourselves – this is the radical empiricism of the early- and mid-twentieth century.

Like VanTil, Feyerabend saw the mind as being tainted, but in his case the matters he identified were wrapped around culture and history. Knowledge was thus a mere social construct. These influences determine what we think and see and conclude. (More on this will follow in the Presentation of Evidence section ahead.) Feyerabend's anarchistic approach becomes the ultimate skepticism, an epistemological pessimism.

Modern empiricism finds its home in David Hume's view of knowledge[134]. But Hume ended up a skeptic, not trusting the mind to assess that knowledge. Kant tried to fix this but his view of personal knowledge ended with an alternative skepticism about our ability to know the real world as we may only know our perceptions. In all of this the certainty of the empiricist was viewed as completely unfounded, even from within their own system. So Feyerabend proposed that we forget these empirical rules, and rules in general, because they (like knowledge) are tainted by culture and history. He questioned, it appears, the foundation for knowledge itself. Yet he opens the door to something better – answering the questions about knowledge by addressing the presuppositions themselves.

What both of these men have done is to demonstrate that there are no "brute facts" to be assessed and that knowledge is indeed not so simple a question. That illusive thing called objectivity can be seen for the illusion it always has been.

The supposed objectivity of the empiricists was, ironically, being put to rest at the same time that their empirical movement was at its height. Theoretical science, physics in particular, was now saying that science was about much more than what is handled and measured. It is also about what can be calculated and estimated. Though empirical work certainly has its place it is now not the sole source of knowledge. Empirical systems have become the support mechanism for model theories and no longer serve at its the foundation.

It is therefore proper and valuable to assess knowledge with some attention given to the influences of that knowledge. Not only may we challenge the theory structures in content but also in historical context. History and culture have fed into it and that history should not be ignored. Here we can include evidence that is ignored (such as the concept of mind), structures that appear to be cultural norms (the influence of the industrial revolution), and even theological construct (the influence of postmillennial theology on the concept of directionality).

Abductive Scientific Theory

Evolutionary theory is first and foremost an explanatory model. It is an attempt to assemble all of the available information on the material world and come to a conclusion about what happened in the past and, at times, to predict what might be in store for the future. It is an explanation of history, natural history in particular.

Studies of history are always abductive. That is, after one begins with a hypothesis and assembles the information according to the hypothesis, one infers a conclusion. There are times when deductive arguments are employed (when dealing with the results of some test or experiment) but these take place in the context of the parent framework of historical abduction. In these cases deduction is subject to the abductive theory employing it.

There is nothing wrong with this method of reaching conclusions. It is done every day in court cases where the evidence is examined according to the hypothesis of the prosecution. The defense attempts to show that there is cause to infer something else. Then the jury comes to the conclusion that one of the two is most correct, whether or not the prosecution's hypothesis is beyond reasonable doubt. The method is sound; what remains is whether there is enough information available so that the jury might draw the desired conclusion. It is the volume and weight of evidence that are at issue in both court cases and in science.

The question at this point is not to doubt the method or to cast a doubt on how the evidence is handled. It is rather to challenge the erroneous message that naturalistic evolutionary theory is somehow deductive. That is, the demand that those scientific models are inescapable and an honest person may arrive at but one conclusion. Evolutionists themselves disagree about what happened in the past, when it happened, and how it happened. They raise disagreements on a regular basis.

A deductive argument is one where the conclusion follows from the premise. A simple example of this is that all bachelors are unmarried. This is true by definition and the conclusion is inescapable. But that does not happen in evolutionary theory. What has seemed to be true is often later shown to be false or at least somewhat in error. For instance, for decades we were shown skeletons of a creature called brontosaurus. But it turns out that the model for the Flintstones' Dino was an error, an assembling of fossils that did not belong together. What was once "fact" is now fantasy.

The Structure of Empirical Criticisms

The initial complaint as noted from Coyne and others often reveals a tendency toward something called *verificationism*.[135] This is the demand that any claim must be verifiable and it is a holdover from the dominant days of empiricism. For instance, "I exist" is testable.[136] Either I exist or I do not exist. Likewise "You exist" is equally verifiable or falsifiable. But to say either "God is love" or "God is hate" is meaningless (to the empiricist) because the statements cannot be verified by any empirically testable system or evidence.

Coyne in the cited work not once employs the term "evolutionary model." Instead the appeal throughout his work is to the data as though data speaks. He takes a generally empirical perspective on the data. As he says on his personal web blog

> And so it is with other things. Can you disprove that I don't have a heart? Of course you can: just do a CAT scan! Can you disprove that I am not married? For all practical purposes, yes: just try to find the records, ask people, or observe me. You won't find any evidence. Can you disprove the notion that fairies live in my garden? Well, not absolutely, but if you never see one, and they have no effects, then you can provisionally conclude that they don't exist.
>
> God is like those fairies. Not only is he a supernatural being who's supposed to exist, but, unlike fairies, a theistic God is supposed to have designated effects on the world. In particular, he's supposed to be omnibenevolent, omnipotent, and omniscient. Some further believe that there is an afterlife in which one goes to either Heaven or Hell, that prayers are answered, that God had a divine son who was resurrected, and so on.

and

> In the case of God, then, the absence of evidence is indeed

evidence for His absence. We can provisionally but confidently say that there's no evidence for a God and therefore reject the notion that He exists. [137]

Here Dr. Coyne has moved from the demand for evidence to claiming that any discussion of God is an adventure in nonsense. God has, to him, the same amount of evidence as does a fairy. But this type of demand has a serious failing. It amounts to, as I see it, an empirical demand from a system that is not an empirical explanation. This is called a categorical error or fallacy.

Recall the various model types. Among them is the explanatory model, which is employed for evaluating history and arriving at the best explanation. Theology is, structurally, an explanatory model. It is (in large part) an historical record of God's involvement with humanity. Theology looks not only at the person of God but also at the works of God through history. This is what we call providence. Things such as fulfilled prophecy along with historical accounts confirm this providence.

In short, the demand for empirical evidence is a misinformed demand. It applies the wrong model to the assertion. It remains with the naturalist to provide an explanation which defeats the historical model presented for the last several thousand as the best explanation of human history.

The Ends of Evolutionary Theory

There are three ideas here that need to be clarified. At the core of evolutionary theory is something called *directionality*. It is implicit, or even explicit, in the idea of common descent that there is some general sense of direction. It is part of the working model that species would move from simple to complex, from small to large, from unintelligent to intelligent, from less able to survive to more fit for survival. Most of the naturalistic origin systems state it clearly and specifically. Mayr accepts the directional concept and even applies this to the rest of the universe (p. 76) when he says:

> Almost everything in the inanimate universe is also evolving, that is, it is changing in a distinctly directional sequence[138].

The problem of directionality is a serious one for many naturalists, and it is admirable that Fodor and Piattelli-Palmarini are willing to tackle it. As they understand the question, the issue of "selection for" is a teleological issue. Through an analysis of the argument structures they conclude "the theory of natural selection cannot predict/explain what traits the creatures in a population are selected-for"[139] so that "the claim that selection is the mechanism of evolution cannot be true."[140] In other words, if there is no *tendency* toward becoming something bigger and better there can be no improvement through evolutionary means. Though this is not the strongest argument that might be made it is a good one. The question of "selected for" deserves an answer.

We earlier discussed Gould's problem with directionality and teleology. The situation goes further with others in the field. Robyn Conder Broyles says

> Evolution works the same way. The rare, random beneficial mutations are kept, because those individuals with such mutations survive better. The detrimental mutations go away in one generation because the individuals with those mutations die quickly. After many generations, the result is a collection of beneficial mutations, without the detrimental ones that arise more frequently but are not preserved. [141]

Broyles' directionality sets the course where detrimental mutations will be quickly eliminated.

Consistent with this, Gould says

> At the species level, not only does each birth of a new individual include novel variation that may be substantial, but the variation also arises in an adaptive context (whereas mutation, the source of variation at the organismic level, will usually be detrimental to the organism). [142]

These statements regarding species improvement via combined mutation and natural selection represent a commitment to directionality. To put it another way, what is detrimental will necessarily be rejected in favor of what is beneficial. There is no real reason why this should be except perhaps for a pre-determined direction.

We may continue to ask the simplest question: Why? Why must detrimental features be eliminated quickly? We know they are not. Why must more beneficial traits develop to enhance survival? They often do not. Who is to say what is beneficial and detrimental and if any species really should survive? These are teleological questions that arise directly from the claims of the theory structure of the evolutionist. Some are willing to take on the issue but few will deal with it as a structural issue that affects the theory itself.

While directionality is itself at issue, it sits in contradiction to Darwin's theory and a principle that is also held by most if not all naturalistic scientists. The principle is that evolution is unguided and without purpose. You may read this in Gould, Coyne, Prothero, and the rest. If evolution is without purpose then why *survival*? Is not survival a purpose, an end, a goal?

Of course they think they really mean to say that there is no external, imposed purpose that comes from some deity. But that does not make the problem go away. They are on one hand asking for a basic materialistic universe and on the other hand asking for some external sense and force to drive the process. It may be impersonal and non-judgmental, but it remains an external purpose none-the-less.

After directionality comes the sticky issue of teleology. Think of it as directionality with a purpose, a driving force that propels it to greater heights. You can imagine how this might raise questions. Why, after all, should a species get better and seek its best, most advantageous position? What reason is there that any species should reproduce to better itself? Needless to say most naturalists will reject this out of hand when confronted with it. But some do not ignore it. Teleology is one implication that is difficult to avoid if directionality is true.

We should not confuse teleology with purpose. That would be a categorical error and would not improve our position. Purpose seems to lead toward a specified end, as we observed with teleology. It is a broad a term and might also be misunderstood to mean some specific purpose. We might best think of teleology as the drive or force behind directionality. For the naturalistic evolutionist directionality is still without purpose or end.

The third and final level is eschatology. Eschatology looks for a final termination point to all of things. Theologians may see a terminus to history but naturalistic evolutionists see no termination to the process of change. This is what is meant by evolution having no purpose. Eschatology thus plays no role in evolutionary studies, though the first two clearly do.

The Extent of Evolutionary Theory

How far does evolutionary theory go? Into what areas of life does it apply? Does it include sociology and psychology, politics and faith, or is it merely a matter of biology and speciation? This question has been around for a long time and the answers to it have proven quite messy.

In his criticism of Jerry Fodor and Massimo Piattelli-Palmarini's (referred to as "F&P") *What Darwin Got Wrong*, Coyne is willing to go take the path of a materialistic and deterministic universe by stating his rejection of randomness in human behavior.

> Fodor has long been an extreme rationalist who believes the mind is a logic machine and that the orderliness of our world must be deducible a priori from elegant laws. It's no surprise, then, that F&P produce a long diatribe against B.F. Skinner's behaviorism, the theory that animals (including humans) initially behave randomly and then repeat those behaviors that get rewarded. In its randomness, messiness and contingency, behaviorism resembles natural selection. And F&P are clearly infuriated by evolutionary psychologists' use of natural selection to explain not only human behavior but the human mind. [143]

There are a couple items to note here. First, F&P see natural selection as a biological process only. They would, as I read it, separate this from learning. Thus from their background they have cause to reject Darwinism (specifically) on account of a psychological evolutionary concern. It is the rejection of mind by reducing it to just another materialistic component that raises concerns.

The idea that randomness can be dismissed at any level leaves few options. One might go with some type of materialistic determinism or fatalism, or perhaps with directionality or teleology. Coyne (elsewhere) chose directionality when he stated that "life on earth evolved gradually, beginning with one primitive species; it then branched out over time, throwing off many new and diverse species—and the process producing the illusion of design in organisms is natural selection."[144]

So what is left if life is besides matter and energy? Not much. Of course there is some attempt to appeal to a middle ground. Though it seems that materialism would certainly yield a meaningless determinism, Coyne stated his understanding that the principle that evolutionary psychology is both messy and random. This means he has room for chance and contingency. But those involved with the mind see something far more complex.

What remains for the naturalistic evolutionist is to draw a line, to make some choice about how biological life progresses. At what point do chance and contingency begin and materialistic determinism, or even fatalism, end? Or at a minimum, what is the scope of their interaction? Is there a compatibilistic option? Where do we find the division between the raw interaction of matter/energy and the awareness of self-awareness? There are many theories about mind, but where mind begins and material interaction ends is an elusive question. There is also no explanatory model. There is only a hypothesis that, as far I can tell, seems untestable.

The Social Context of Evolutionary Theory

The material in this segment is intended to provide some historical context to evolutionary theory. It amounts to an application of evolutionary theory regarding human behavior to the theory itself. If evolutionary theory is to serve as an explanation of social constructs such as human interaction (notably because this is a type of adaptation), then the principle might be returned to the scientist's context in similar fashion. In other words, are the conclusions of theorists about social constructs themselves engaging in a meaningless, evolved social constructs?

The nineteenth century, and even the two centuries which preceded it, were times of great human progress. Philosophy and music were reaching a peak. The economies of Europe were flourishing on account of a combination of secular Renaissance entrepreneurship (imperialism) accompanied the German Protestant (Reformation) work ethic. It was the people of the Renaissance who funded the era of exploration and it was the work ethic of the Reformation which gave the commoner, with the transition from feudalism to an open economy with this new thing called the "middle class" (the business class, the bourgeois), cause to get up and move and build and create by himself and for himself. As Roman Christendom struggled against skepticism both individualism and that new entity called the nation-state expanded.

Protestant evangelicalism saw the birth of the missions movement of the 19th century. Missionaries were sent by the hundreds around the world and evangelicalism blossomed globally as preparations were made for the coming Kingdom of God. This was the fruit of the (post)millennial fever of the era. God's kingdom needed to be established in anticipation of His soon return.

There were other millennial movements of the era as well. This period saw the blossoming of premillennialism and with it dispensationalism, the rise of the Seventh Day movement as well as Jehovah's Witnesses and Mormonism, all attempting to say something about the coming Kingdom of God. In no other era could they have come to be. Yet postmillennialism was the dominant theological construct of the era.

The mechanics of this modern era brought heretofore-unknown levels of leisure to many. It appeared that the world was getting better. As a result postmillennial theology grew in both popularity and influence. The secular variant of postmillennialism, a variation of liberalism known as the progressive movement, was also gaining significant momentum.

The idea of progress was "in the air" as they say. Progress in the human condition was understood as the norm. A better world was to be expected. Why? Because it was what everyone was observing. So, it is my contention that Darwin could not escape this, either in the secular progressive context or in the theological context. What was Darwin explaining? He was explaining the progress of species. Directionality was implicit; it was the subtext. Directionality in evolutionary theory, it seems, may not be as scientific as it is sociological and even theological. Likewise Darwin's evolutionary theory (including his near predecessors) could also not have occurred at any other time in history.

Paul Nelson's discussion of the presence of theology in Darwin's model proceeds this way

> But what we see as *the evidence* for evolution exists against an epistemological backdrop where theology of one form or another has always been present. The panda's thumb is a sign of history - i.e., of descent - only when one is certain that "a sensible God" (Gould 1980, p. 20) would not stoop directly to contrive such oddities.[145]

What Nelson is describing here is that evolutionary theory is often a reaction and not immediately science. Theology is inserted into the discussion by the scientist themselves since they are seeking more than just an alternative explanation. They seek a replacement theory. The skepticism of our Rationalist era is not so much about finding what is true as it is about questioning the past. It is in this context that Gould questioned what a supposedly "sensible" God seemingly should have done. The discussion about science is not always about science.

For example, the imperfection argument is raised as an apologetic to support the idea naturalistic origins. It does not simply promote evolutionary concepts but is used to counter theological claims regarding creation. Nelson goes further to make this point:

Or one might try to justify theological arguments for evolutionary science pragmatically, as devices for shutting up the creationists. The arguments are indeed theological, this justification holds, but only because of their peculiar context. The arguments take the logical form of *reductio ad absurdum*, where one assumes the truth of an opponent's premises provisionally to derive a contradiction from them. Terms like "God" and "the Creator" appear in the arguments because they were introduced first by creationists, in *their* arguments, into a cultural debate about the truth of evolution. "God" is the principal cause invoked in non-scientific theories, and such theories do have genuine observational consequences. "If theology presumes to speak of the natural, material world," argues evolutionary biologist Bruce Naylor (1982, p. 94), "its statements become open to scientific examination and potential falsification." The panda's thumb, in other words, can be stuck in the eye of the creationists. As a polemical tool, therefore, theology is useful. But evolutionary theory *as a natural science* claims nothing for itself theologically. When the debate is over, the theology, borrowed for the evening's *reductio*, goes into the trash bin with the folded programs and coffee cups.

Or more briefly, why even discuss theology at all if theology is not part of the theory? But at this point the discussions around creation are more apologetic and not part of the structure of the theory. Yet they are tied very closely as Nelson pursues further in Darwin's language.

But note again that little indicates Darwin ever rejected the deep presuppositions that he inherited from English natural theology, namely, perfection as an observable quality of organic design, and the conventional conception of the nature (if not the actions) of the creator. Indeed, a close reading of the Notebooks would suggest that Darwin saw his theory as providing a more sublime conception of the actions of the creator (see, for instance, D 36: "What a magnificent view one can take of the world ... "). *Darwin employed a particular conception of God to judge theories of God's creative activity.* Otherwise, why should the multiple creations scornfully derided in D 37 as a "long succession of vile Molluscous animals" be beneath the

85

"dignity" of God? Cornell (1987, p. 397) argues, of this and other passages from later notebooks:

> As always, Darwin's idea of "perfection" refers to the nice relationship of organisms to their physical surroundings. But it also refers to the overall design of the world, from a divine viewpoint. ... Darwin's sense of a comprehensive system, the invocation of divine perfection, and his new theory are thus all closely related.

And, as Brooke (1985, p. 46) argues:

> The fact is that there are several entries in the transmutation notebooks which indicate that Darwin was discovering a philosophy of nature which he genuinely believed conferred a new grandeur on the deity, despite - or rather because of - the fact that it superseded Paley.

We are now a step closer. Theology was and is countered as both an apologetic and influence on the theory's development and can be shown to have some real influence as well. But was there any actual positive use of theology in the theory structures of Darwin? Stephen Dilley provides us with a discussion of Darwin's positive theology in *Origin of Species*. As stated before, directionality is dependent upon a theological construct.

It remains difficult to avoid Darwin's theological construct. It appears that not only is the language present, but its presence informs evolutionary theory.

> While current evolutionists may be indifferent or opposed to Darwin's theology, their use of the imperfection and homology arguments for evolution presupposes the intelligibility of notions rooted in Darwin's theological metaphysics: perfection as an observable quality of organic design, and the intuition at the heart of Darwin's metaphysics - that a rational and benevolent God would have created an organic world different from the one we observe. Both continue to inform evolutionary theory.[146]

That is, the nature of the naturalistic apologetic works with certain assumptions that were drawn from a theological metaphysic. Given the time and place in history it is understandable.

Dilley thus identifies directionality as part of an Enlightenment style theology. As Darwin wrote near the finale of the *Origin of Species* that

> Authors of the highest eminence seem to be fully

satisfied with the view that each species has been independently created. To my mind it accords better with what we know of the laws impressed on matter by the Creator, that the production and extinction of the past and present inhabitants of the world should have been due to secondary causes, like those determining the birth and death of the individual.

In this passage, Darwin compared two theories in light of a claim about natural laws. More exactly, he compared special creation and evolutionary theory to a background claim – what 'we' already 'know' about the laws of nature – in order to assess which theory best accorded with this background knowledge. Darwin reasoned that knowledge about the laws of nature favoured the 'production and extinction' of flora and fauna by 'secondary causes' rather than by independent acts of creation. This claim implied that the laws were unbroken – otherwise they could not favour purely secondary causes (and, hence, descent with modification) rather than miraculous causes (used in special creation).[147]

and

Theology provided the edge. Laws were 'impressed laws on matter by the Creator'. The language of 'impressed' laws, I would argue, suggests a picture of a Creator who, having once implemented these laws, allowed nature to act only by secondary causes. Consider, first, Darwin's endorsement of this view elsewhere. In his autobiography (which he originally intended only for his family), Darwin affirmed the dichotomy between laws and miracles: 'the more we know of the fixed laws of nature', he wrote, 'the more incredible do miracles become'. He also claimed that nature operates by laws alone: 'Everything in nature is the result of fixed laws'. Darwin's early notebooks reflected the same sentiments, privately endorsing divine creation by law as 'far grander' than specific instances of creation by miracle, which were 'beneath the dignity of him, who is supposed to have said let there be light & there was light'. And, in his 1844 manuscript, he added that 'laws capable of creating individual organisms . . . should exalt our notion of the power of the omniscient Creator'. Thus, in writings before and after the Origin, Darwin consistently rejected miracles and instead favoured unbroken natural law.

Such Enlightenment theology reads like an expression of deism – that the creator is out there, established fixed laws, and after that was uninvolved. Though Darwin's passages are held by some to be controversial at best, it would be fair to conclude that an individual remark might be disregarded but a series of remarks reflects the mindset of the era and language that he was predisposed to employ.

We have taken another step closer to theology informing scientific theory. Now we find that the theology of the day had a direct influencer on the structure of his theory. But as to the content of the theory structure we take just one last step. Dilley goes further with this conclusion:

> Arguably, at a deeper level, Darwin's theological struggles with suffering did not simply provide evidence for his theory but also shaped its content as well. Recall the quote from the Origin (above), in which Darwin stated that instances of natural suffering are 'small consequences of one general law, leading to the advancement of all organic beings, namely, multiply, vary, let the strongest live and the weakest die'. In this passage, he characterized the work of variation and selection as leading to 'the advancement of all organic beings'. Natural suffering, apparently, was acceptable collateral damage in a process that ultimately produced a higher, better outcome. God designed the laws of nature in keeping with his moral nature: these laws led to progress so that the final end justified any suffering along the way. Natural selection thus insulated God from the suffering so troubling to special creation; it exonerated God from direct responsibility for natural suffering while also ensuring that the struggle for existence led to a morally acceptable end. Thus Darwin's theology sanctioned his theory. The belief in a distant, yet moral, God required a means of creation that could account for the presence of natural suffering in a manner consistent with God's character. Variation and selection, with their progressive element, provided just that.

So while Darwin's conflict was against some presumed perfectionism in creation what he proposed was a remote deity creating this process of natural selection for "the advancement of all organic beings." Advancement is directionality and advancement appears to have been directly informed by theology. So while Darwin was opposing certain assumptions of the theology of his day he did at the same time make use of that theology to advance his theory and he did so in the theory structure itself.

The Fruit of Evolutionary Theory

For those of us who hold to moral standards based on the Bible it is possible to say that evolutionary theory has given us both good and bad. On the good side we might acknowledge that medical advances involve a study of the genetic behavior of viruses, et al, from a neo-Darwinian framework. Probably the most popular of current observations is with respect to HIV mutations[148]. We may, and ought, acknowledge the fruit even if we would disagree with the assumption of naturalism.

The negative fruit of naturalism is generally ignored and even argued against by the naturalist. Of course not all agree with the immorality of the situation. (One should also not miss the appeal to Christian morality for convenience when the ugliness of a position becomes clear.) What is this fruit? Jacques Barzun, most certainly not a theist, raised the specter of this relationship between ethics and materialism many years ago.

> The reply is simple: the evil world we live in is not a world which has been denied access to the science of Darwin and Marx and the theories and art of Wagner. Had their answers truly solved the riddle of the Sphinx, no obscurantism could subsist, for we are animate by – I will not say, the precise ideas of the three materialists – but surely by their deeper spirit, their faith in matter, their love of system, their abstract scientism, and their one-sided interpretation of Nature:
>
>> Thus, from the war of nature, from famine and death, the most exalted object of which we are capable of conceiving, namely, the production of the higher animals, directly follows. There is grandeur in this view of life. ...
>
> This is not Mussolini speaking, but Darwin, and his voice re-echoes in our ears:

> War is not in contrast to peace, but simply another form of expression of the uninterrupted battle of nations and men. It is an expression of the highest and best in manhood.

> The last is the comment of Dr. Robert Ley, head of the Nazi Labor Front, on the war of 1940.[149]

What Barzun is saying, quite clearly, is that although he does not see Darwin (with Marx and Wagner) as the direct and immediate causes of the national sins of the 20th century, he sees what proceeded from them – their spirit and perspective. Ideas do not sit in isolation and their influence is felt. Though the fingerprints of Darwin may not be found in these events the naturalism and materialism promoted by them is clearly evident.

We dare not ignore the full scope of the eugenics movement and their effort to escape this relationship. Based on evolutionary theory and coupled with progressive politics, the goal was and is to create a better, a stronger human race. Eliminate the inferior and what is left is stronger. It will survive better as it is more suited for survival. It will conquer, as noted above.

Nobody has a problem assigning eugenics to the Third Reich. The Nazis were blatant about it. But at the same time eugenics was also strong in other parts of the West and it remains so today.[150] Unfortunately we are so close to it that we may fail to see the forest for the trees. At the same time as the Third Reich, Margaret Sanger (consultant to the Reich) suggested that we

> Give dysgenic groups [people with "bad genes"] in our population their choice of segregation or [compulsory] sterilization.[151]

To deal with a population group that she wished to reduce she suggested that

> We should hire three or four colored ministers, preferably with social-service backgrounds, and with engaging personalities. The most successful educational approach to the Negro is through a religious appeal. We don't want the word to go out that we want to exterminate the Negro population, and the minister is the man who can straighten out that idea if it ever occurs to any of their more rebellious members.[152]

Her position is further enhanced in her dim view of life for children already born into a harsh world.

> It is these conditions that produce the 3,000,000 child laborers of the United States; child slaves who undergo hardships that blight them physically and mentally, leaving them fit only to produce human beings whose deficiencies and misfortunes will exceed their own.[153]

Her proposals for sterilization, and elsewhere equivocation on infanticide and abortion, are the tip of the iceberg. The issue went beyond Sanger's efforts. Bennett Gershman, professor at Pace, said these things about the progress of the eugenics movement in the U.S. in the 20th century.

> This insidious quotation ["Three Generations of Imbeciles Are Enough"] was the tag line in Supreme Court Justice Oliver Wendell Holmes's infamous opinion in the 1927 case of *Buck v. Bell*, which legitimized the eugenics movement in the United States, involving forced sterilizations on a massive scale of persons deemed "socially inadequate" to bear children. Unfit persons included the poor, illiterate, blind, deaf, deformed, diseased, orphans, "ne'er-do-wells," homeless, tramps, and paupers. This evil experiment in social engineering was intended to purify the white race; the program served as a model for the Nazi racial laws and Hitler's "final solution" for exterminating the Jews. Holmes rejected Carrie Buck's plea to prevent her court-ordered salpingectomy to make her sterile. The eugenicist for the state of Virginia claimed she was "feeble-minded," a controversial and indeed fraudulent label used to justify the procedure. In fact, as disclosed many years later, Carrie Buck was a poor 18-year-old woman of normal intelligence who was ordered sterilized to hide the shame of her pregnancy after being raped by a relative of her foster family.

And

> How did this crazy experiment in human genetic engineering in the U.S. become so influential and with such devastating consequences? Wealthy businessmen, powerful politicians, and a phony intelligentsia, steeped in the

fashionable Darwinian theory of the "survival of the fittest," drove the movement, whose goal was to purify the white race by preventing the propagation of those persons considered biologically unfit to bear children and encouraging procreation only by those considered worthy. More than 31 states launched government-run eugenics programs, and over 65,000 persons were sterilized. A "Model Sterilization Act" was proposed by officials at the central eugenics agency - with the Orwellian title "Eugenics Record Office" — which became the model for states as well as for the racial laws in Nazi Germany. The eugenics movement lists as one of its greatest triumphs the severe restriction on immigration, with national quotas that discriminated against those considered mentally inadequate.[154]

As much as our society is today unaware of these practices and wrongs, the naturalistic evolutionist community too often sits in denial and argues that Darwin had absolutely nothing to do these concerns[155]. Fortunately men like Gershman, no friend of the pro-life movement, are able to draw the proper historical connection.

One may also explore the justification of slavery via amoral implications of sociobiology and evolutionary psychology.

The term 'sociobiology' was introduced in E. O. Wilson's *Sociobiology: The New Synthesis (*1975) as the application of evolutionary theory to social behavior. Sociobiologists claim that many social behaviors have been shaped by natural selection for reproductive success, and they attempt to reconstruct the evolutionary histories of particular behaviors or behavioral strategies.[156]

This is the field to which Coyne was referring in his criticism of Fodor and Piattelli-Palmarini. The idea goes further than just making scientific claims and makes certain assertions regarding social matters. At times it becomes, or approaches, dangerous in its implications.

Sociobiology has been less successful in its application to human behavior than in its application to non-human systems. According to many critics of human sociobiology, standard sociobiological models are inadequate to account for human behavior, because they ignore the contributions of the mind and culture. A second criticism concerns genetic determinism, the view that many social behaviors are genetically fixed. Critics of sociobiology often complain that its reliance on genetic determinism, especially of human behavior, provides tacit approval of the status quo. If male aggression is genetically fixed and reproductively advantageous, critics argue, then male aggression seems to be a biological reality (and, perhaps, a biological 'good') about which we have little control. This seems to be both politically dangerous and scientifically implausible.

The syllogism of the evolution changes here. Darwin began with Adaption + some unknown influence to yield change. The neo-Darwinist altered it to be Random Genetic Change + Adaptation to produce change. This has been updated and clarified to occur within the context of a group rather than simply the individual as the point where change occurs. Now, if a change in one group is superior to the change in another group, is it not an evolutionary advantage that the group with the more advantageous traits would gain superiority over those groups with less-advantageous traits? Is this not also the case once one accepts the community nature of neo-Darwinism even without the formal descriptions provided by sociobiology?

The moral questions follow regarding social evolution. For instance, does allowing the strong to subdue the weak enhance the human species? Do slavery and aggressive warfare find justification in Darwinian and neo-Darwinian theory? If there is no basis morality, can any claim to "just" or "unjust" ever be made on what happens between social groups or nations? Only a functional morality may answer these issues.

The absence of morality in the mind of the empiricist declares the questions nonsense for they are non-empirical assertions. The consistent empiricist, then, bears the burden of these consequences. Of course most students have not given consideration to the problem. The questions, let alone the potential answers, have not even been broached. It's difficult if not impossible to answer a question that you've never been asked.

On the Presentation of Evidence

One might be tempted to believe that, once all the facts are laid out, that there would be only one conclusion (or at worst a limited set of conclusions) which should be obvious to all thinking, all intelligent persons. Take a deck of cards, spread them out, and see four suits of thirteen cards each, plus a pair of jokers. Fifty-four cards in all. Nobody with any sense would say anything else, at least not under normal circumstances. But information is not always quite so clear and the presentation is not always without prejudice. Not all facts are as plain as the counting of cards in a deck. Models are not generally constructed to deal with a full set of evidence but also with a partial set and so to create a functioning model to explain the data.

The model that describes or proposes a materialistic and mechanistic world may not be so plain. One question to consider is how such a mechanistic view of the world could have come into being. With it comes the corollary question of how progress, i.e., directionality, could have been accomplished in a mechanically determined and closed system.

To the first question, and still related to the second, we take an inference from history. Consider the world in which Darwin (as well as Wallace and other influences such as Malthus) lived. This was the world of the industrial revolution. Jacques Barzun puts it like this:

> It is true that we are not bound to guide our daily life by the rules or assumptions that work in the laboratory, but the historical fact is that this is what Western man since the late nineteenth century has tried to do. Mechanism and materialism worked so beautifully — for a time — that even now we can sympathize with the enthusiasm of those who by its aid bequeathed to us the industrial world we live in. The misfortune was that when mechanism began to be questioned, for scientific reasons, the general public had become persuaded of its absolute truth; it could think in no other terms and it felt that all other views were simply "prescientific."[157]

During that century (the 19th) the moral economy of the world was also growing rapidly. Social organizations were caring for animals as well as humans. Slavery and indentured service in the United States were brought to an end. The new "social gospel" was working to improve the lot of the poor. Evangelicals were engaged to the same degree as the theological modernists and liberals. It appeared that world peace was coming soon. This mechanical view of the world was just "in the air." Though today we might wonder about that perspective, as Barzun points out, the attitude of what is not "scientific" persists today.

Now, we know that such a worldview did not occur in a vacuum. The idea of "science versus faith" had already become entrenched by the Rationalist movement over the roughly four centuries preceding that time. This new peace movement was accompanied by the idea that religious movements were the source for violence and myth. It was their apparent failure which hurt humanity and which Reason alone might solve.[158]

At the same time as the industrial revolution was a parallel change in theology. The 19th century bore witness to the dominance of postmillennial theology. The church was at work establishing the kingdom of God, preparing His world for His return. During this period the evangelical world moved forward with a massive missions program and even liberal theologians were preparing a better world through social action, as previously noted.

Darwin was also writing in a theological world. He knew theology and also appealed to theology in his work. Dilley notes, for instance, that Darwin made an appeal to Natural Law much the same as is done in the Declaration of Independence. He also says:

> More directly, Babbage, Herschel and Whewell used the identical word – 'impressed' – in order to express the same idea of God inscribing matter with enduring, lawful qualities. For example, in the Preliminary Discourse, Herschel wrote of the raw materials of the universe: 'by creating them . . . endued with certain fixed qualities and powers, he has impressed them in their origin with the spirit . . . of his law, and made all their subsequent combinations and relations inevitable consequences of this first impression'. And Whewell declared in his Bridgewater Treatise,
>
> > God is the author and governor of the universe through the laws which he has given to its parts, the properties which he has impressed upon its constituent elements: these laws and properties are, as we have already said, the instruments with which he works . . . through these attributes thus exercised, the Creator of all, shapes, moves, sustains and guides the visible creation.
>
> This statement occurs in the same paragraph as the claim Darwin borrowed from Whewell as an epigraph to the Origin, which stated that God does not work by discrete miracles but rather 'by the establishment of general laws'. Whewell went on in the next two paragraphs to quote Francis Bacon and Herschel to the same effect, citing the Herschel quote above. [159]

What seems evident is that Darwin's argument, at least in part, required something external to drive it. It required something to set its course, its direction. His mechanical naturalism required a law to enforce directionality, to produce better species, a better humanity, and a better world of living organisms. One might infer that Darwin's closed system, into which no deity may presume, is only closed for the convenience of his theory. Here it appears little different from modern theistic evolution, as least in this core assumption. Darwin was clearly not working outside the confines of the world he lived in. One's culture and surroundings often define one's beliefs and even one's science more than might be at first thought.

As a defense against this some of Darwin's proponents might suggest that if the logic and science appear sound it would make no difference where and when his efforts occurred. To answer this we would raise another question: Could Darwin's (and his predecessors') ideas have occurred at any other time in history. A negative response here is certainly valid. His theories operated in a mechanical world of assumed social progress, both political and theological. One might even say that the conditions were fine-tuned for his work.

Responding to the Imperfection Argument

The imperfection argument can be quite compelling. It asks why imperfections exist when design ought to have created something optimum rather than detrimental. The most frequently cited characteristic is the laryngeal nerve. It runs a circuitous path, going from the brain to and around the heart area and then back to the larynx for speech production. According to the evolutionist this creates a vulnerability that an intelligent creator would not have introduced.

There are several ways to respond to this. Here I would prefer to appeal to the principles of evolutionary science as they have been presented. The query to pose in return is: Why do some traits persist when they are not needed? Why do they remain even when they are clearly harmful? If evolution is to produce better beings and if negative and harmful traits are removed quickly from any species, why is it that burdensome traits remain? Should they not have disappeared quickly in favor of traits developed for better survival? It is the position of many evolutionists that deadly and harmful traits should disappear quite quickly. One should question the character of the model if this assumption is to be maintained.

Of course not all accept the idea that all negative traits will disappear. It's really not necessary that they do. What is necessary is that new traits be useful so that the creature can adapt, or that the creature's adaptations bring about change, some of which is advantageous. But even that position raises an issue.

Also, the simple question might be asked: How do you know? Earlier imperfection arguments revolved around the appendix' apparent lack of function and later the concept of "junk DNA" as a mere vestige of past development. Today we know that these claims were in error. We know much more about the usefulness of each. It is a fair assertion that they are engaging in what we might call "evolution of the gaps" argument.

This is a variation on the old "God of the gaps" argument where the Christian would be labeled ignorant for making the claim that "God did it" on the simple basis that something did not yet have an explanation. But substitute evolution for God in the equation and you get a variation on the argument from a naturalistic frame of reference. Just because we do not know something does not mean that it can be attributed to some obtuse model claim that seems to work but has no evidence. That is, evolution has become an orthodox position, an answer that cannot be questioned. Once that position is held the acolyte is free to claim "evolution did it" whenever evidence is lacking.

Another way to respond is to take the positive side of the argument. That is, if this is a traceable evolutionary feature then perhaps we need to explore and examine – trace back – other features of seemingly greater significance. For instance significant characteristic of humans versus apes is that our hips are turned inward, thus allowing our upright position. We also maintain a larger cranial capacity than do the apes. So what is in between? Is there any creature with a large cranium but turned out hips? So far, no. Is there record of a creature with lower cranial capacity but turned in hips? Not yet. Australopithecus Afarensis (aka "Lucy") has, as Gould states, "an apelike palate, its human upright stance, and a cranial capacity larger than any ape's of the same body size, but a full 1,000 cubic centimeters below ours."[160]

Lucy is a fascinating study in how the evolutionary models are built. The significant question I would pose here is to ask how old Lucy is? Most dating is stated as 2.6 to 3.2Ma (million years). But from Greece has also come the estimated 10Ma Ouranopithecus macedoniesis. This is seen by some to be a primate and also a relative of Lucy's ancestors.

Now the dilemma: What of the Yucatan impact of 65M years ago? The assumption is that after that event the only mammals existing were small rodent-type creatures. We have remaining 50 - 55M years to accomplish all of the changes needed to transform from some mere rodent to today's human species. That means we need a significant number of changes in a relatively short period of time. The implication here is one of time and speed.

When we start talking about a million years the scientist will say that a million is a long time. But it is during that million that multiple changes need to take place. Some of these changes will be beneficial and others detrimental. And since it is not a simple black-and-white distinction, some will be a little detrimental and may be retained without doing serious harm while others with minimal benefit will also come and go.

Modern recorded history has covered roughly five thousand years. In that time humans have not changed. That period of time is 1/200 of a million. If many changes are to occur within a million years then a few changes might be predictable within a few thousand years. Some of these would be detrimental and so would be lost. Others would be helpful and would increase human survivability.

We know a lot of human history. We have human history in written form covering more than four thousand years. In that humans have gone through no changes that would redefine the species. There have been adaptations such as hemoglobin C for survival against malaria. But in evolutionary terms even this does not count as an adaptation as it did not come about in response to the malaria condition and so become universal in the species. It appears to be a mere coincidence. The implication here is staggering in terms of numbers. We have apparently observed nothing in terms of biological change in humans in over these few thousand years. Nothing, that is, which would alter and redefine the species.

It appears that the thread for tracing back features is somewhat broken by major historic events. If as Gould states this laryngeal nerve has its history in reptiles and amphibians, it would have been passed through that small number of mammalian creatures living after the Yucatan impact. But do we have evidence in the fossil record of this? (Is this not an argument from morphology as Darwin did rather than an argument from an evidence trail?) The simple fact is that there is no fossil record of such soft material. To argue from current data and infer a past condition based on similarity rings more of the method of Lamarck or Darwin looking at similar embryo shapes and arguing, not from empirical evidence, but from appearance. Inference without evidence amounts to an empty assertion. In the language of the new atheists that which can be proposed without evidence may be properly dismissed without evidence.

Also important here is the principle of sequence in adaption. It is not fitness that creates change. That's the error of Lamarck. The neo-Darwinian principle is that change creates fitness. Genetics creates change and the beneficial changes produce better survivability and strength. We observe no significant change in anything over the last several millennia. So either the process has reached stasis, a condition which some claim as part of the whole process, or the assumptions about constant change might be in error.

Here we should be careful. There are some animals which we treat as separate species but which do reproduce successfully. There is the horse and donkey which will create a mule, but which cannot produce past one additional generation. Then there are the lions and tigers that still can cross, as well as the llama and alpaca among the camelids, which also have the capacity to reproduce.

Though this sounds a little disjointed there is an important principle to keep in mind: The over-generalizations of naturalistic evolution may not be so clean-cut. Negative traits may and do persist through generations (e.g., sickle cell). Species may also develop without isolation (e.g., llama and alpaca). What appear to be different species may be simple variations within a species (e.g., all domestic dogs and wolves). One must question if one million years is enough to accomplish any substantive change.

But if we are not seeing these changes then we run into another problem. The evolutionist doesn't like the idea of saltationism[161] (the root is the Latin word for "step" or "leap"), the idea that great changes happened abruptly and to an extreme degree. No lizard egg will suddenly yield a bird. Such a giant leap is preposterous to everyone today. But if the changes came in close groupings then something similar (short leaps rather than long leaps) seems the only option. By that I mean that feathers had to come into being at some point and their existence had to serve a survival purpose. Some gene regulator and accumulation of traits had to be turned on at some point.

Of course a creature with feathers is not that new, fantastic creature out of nowhere. Such is the dream of the critic who has not read the material. But it is, as we have noted from genetics, that changes may be significant. A new feature might appear over just a few generations, maybe even a timeframe as short as a century or a millennia. So it might be proper to call this a "soft" saltationism. Perhaps "saltationism-lite" might suffice.[162] Something as significant as the addition of a new feature in DNA and other genetic material would certainly count to this principle.

Consider the "'Game Changer' in evolutionary science from S. African bones" discussed earlier. Here the concern was around a discussion of the cranial capacity. But again the question of how accurately the model might predict what would fill in the historical data comes into question.

On the one hand the evolutionist says that the information was a total surprise. But the statement is made that this is exactly what evolution would have predicted. Now one cannot have it both ways. Either the situation was predictable in which case there should have been no surprise or there was something unexpected because it was not predicted. Statements like this ought to leave the reader wondering whether or not the model in play was actually predicting something or whether the supposed prediction was nothing more than the wishful thinking of the evolutionist.

There are times when this happens in other fields. One is often tempted to change the story when it seems convenient. Situations such as this have been noted in the social sciences. As an example, it has been noted regarding Marxism; when the system fails the proponents always have an excuse[163]. It seems failure is always someone else's fault. Things weren't done right. Implementation was not what it needed to be. There are always excuses available to those who will not acknowledge that their model does not answer the question, but who stubbornly refuse to abandon a failing model.

In the end, the imperfection argument falls short because of a lack of knowledge about a feature, an appeal to ignorance ("evolution of the gaps"), and a failure to account for enough time to answer the questions of measured change.

Some General Analysis

This mix of argument methodologies opens up some questions. The issue that I take with these positions is that the argument becomes so incoherent and so broad as to become unfalsifiable. When the whole claimed "model" becomes little more than a mixture of convenient arguments then it can be called, not only unfalsifiable, but also irrational, at least as an argument goes. (By irrational I do not mean incomprehensible, but instead convoluted or unsound.)

That does not mean that all of the conclusions in naturalistic science and evolutionary theory are wrong. If that were the case then all lab and historical studies would need to come to be terminated because they would all be rejected out of hand. What it does mean, though, is that particular conclusions become less trustworthy as they are being supported by arguments that are so weak that they can be rejected as not being sound even though fruitful. A systematic where the arguments alternate between the presupposition and the evidence appear by their nature to be viciously circular arguments. The end product of many of the arguments of the naturalist is an elegant incoherence.

Concluding Remarks

All this leaves the Christian with something important to keep in mind. What is called science is often philosophy and sometimes even theology. There have been many who rejected the Darwinian and neo-Darwinian model because of their respective errors and have done so having both scientific and philosophical justifications. Though the demand for orthodoxy in the Darwinian world is often heavy-handed a clear understanding of the subject along with a fair representation of the facts will help establish the position of the Christian as both skilled and expert. The wisdom required is both intelligent and uncompromising. We may be the ones best capable of fixing what is broken.

END NOTES

1 William Cavanaugh discusses the historical issues in *The Myth of Religious Violence: Secular Ideology and the Roots of Modern Conflict*. The continuation of this division in philosophy is further documented by John F. Hurst in *History of Rationalism, Embracing a Survey of the Present State of Protestant Theology*.

2 Coyne, Jerry A., *Why Evolution is True*, Viking, 2009, p. 92

3 Prothero, Donald R., *Evolution: What the Fossils Say and Why It Matters*, p. 15

4 The first popular removal from a position was Forest Mims, III. This situation is well-documented. http://www.forrestmims.org/scientificamerican.html

5 It is here that we end up doing exactly what is expected. If we treat our faith in God as something which has no place in the world of science then we give effective acknowledgement to the secular idea of the two being separate. This is the point where our understanding of the scope of the Christian faith needs to be re-evaluated and seen in its full breadth and scope.

6 This is the only location where I have seen a replica of the "Lucy" skeleton.

7 An apologetic which takes a defensive posture and only attempts to deflect an enemy or opponent seems destined to defeat. In contrast to that, an apologetic which sets its sights on victory over an opponent's foundations has more opportunity for success because it has victory as its goal.

8 The brontosaurus never existed. It was created by an error, or at least that's the story at one point in time. The correct creature was apparently an Apatosaurus. http://www.npr.org/2012/12/09/166665795/forget-extinct-the-brontosaurus-never-even-existed

9 There is a spiritual component of the Bible that is not being ignored. The spiritual character expresses itself through the history of God's redemption relationship with humanity, That is why Christianity can confront cults such as Mormonism which do not have a verifiable historical account. Because the Bible is history your Bibles contains maps of real places and records of real events.

10 There are two general approaches to describe evolution. The first is the level at which it takes place. These levels include biochemical evolution within cells, micro evolution within a species, and macro evolution which creates species (either in the general population or in subgroups). Along with this are the types of evolution. The process is postulated to occur on account of adaptation, genetic change, or some combination of both. In addition to these there are questions of rate. Does it occur at an infinitely slow rate (uniformitarian gradualism), with some periods of acceleration (phyletic gradualism), or with rapid changes in relatively short periods of time (punctuated equilibrium). Each of these several understandings has created various schools of thought, like denominations, within the scientific community.

11 This question challenges the classic Darwinian, the older neo-Darwinian, and the more recent neo-Darwinian on how the model might actually accomplish anything.

12 Alternatively one might say that the abstract concept of justice is universal while the execution of justice varies between cultures. To this I would respond that such a broad view of justice leaves it without foundation, an abstraction without particular definition. It is here that the term loses meaning and is akin

to saying that all "religion" is equal as they seemingly share a same higher good. We know that the latter is untrue and the former is equally untrue on the same basis.

13 Presuppositions are the most basic principles that one holds, whether stated or not. They are what drives our behavior and thought. From Van Til's Christian Apologetics, p. 5 and p. 98: "Presuppositional apologetics asks that we but recognize that all ideas and arguments come within a basic arrangement, a framework within which they make sense. (p. 5)

That is, in the last analysis, the question as to what are one's presuppositions. When a man became a sinner, he made of himself instead of God the ultimate or final reference point. And it is precisely this presupposition, as it controls without exception all forms of non-Christian philosophy, that must be brought into question. If this presupposition is left unquestioned in any field, all the facts and arguments presented to the unbeliever will be made over by him according to this pattern. (p. 98)"

Assumptions are the mechanisms that give the presuppositions legs. The Christian operates with the presupposition that God is (Hebrews 11:6) and the assumption that God created (Genesis 1:1). The Christian views the existence and creative efforts of God as much more than a philosophical construct. But that does not mean the structure of the principle is any different, and that is the only point here – this is a matter of how the discussion is structured, not the truths behind the argument.

14 This is the question of Cornelius VanTil, Alvin Plantinga, and Paul Feyerabend. For VanTil (*Christian Apologetics*) the issue is that our knowledge, as well as our ability to know, is tainted by our fallenness. Plantinga (*Warrant and Proper Function*) raises the same question regarding whether or not our faculties have the capacity to function properly if they are the product of something meaningless. Feyerabend's approach (*Against Method*) is similar. Though he does not accept the Christian concept of human fallenness he does challenge our assumptions of objectivity and whether or not information may ever be treated as brute facts because we bring to the table more than we realize. This includes our personal history, our culture, and so forth. For Feyerabend there is a pessimism about any ability to create any objective test.

15 This is Hume's problem of induction. Because so much knowledge is gained inductively there is always a level of uncertainty, even if that level is minimal. It seems, unfortunately, that the pessimism of our day has made this uncertainty seem more significant than it ought be.

16 Physicalism is a limitation placed on inquiry. It is the demand that the processes we employ are only dealing with the physical world. It is a restriction on the scope of the methodology and is not a statement regarding the assumptions brought to the inquiry. Naturalism comes in two segments. Methodological naturalism, sometimes confused with physicalism, includes the assumption that only the physical exists. Metaphysical naturalism then carries this position a step further by declaring that there is nothing but the physical. Physicalism is also core to the Received View (Suppe, 1977, p. 15) as the RV is an inherently empirical system.

17 Suppe, Frederick, *The Structure of Scientific Theories*, 1977, University of Illinois Press, pp 50-51.

18 Ibid, p. 13, 15. "Since metaphysical entities are not phenomenal or observational entities, the terms used to describe them cannot be observational terms, and so must be theoretical terms. But, theoretical terms are allowed only if they can be provided with correspondence rules which give them an explicit

phenomenal definition, and so the objectionable metaphysical entities cannot be introduced into scientific theories."

"... the Received view embodies physicalism as a doctrine of perceptual knowledge." This premise will have significant impact on Intelligent design and irreducible complexity and their relationship to scientific theory. But the damage done is not serious as I intend to show that ID/IC can be reformulated.

19 Popper, Karl, *Conjectures and Refutations: The Growth of Scientific Knowledge*, Routledge and Keagan Paul, 1963, pp. 33-39

20 Suppe, Frederick, *The Structure of Scientific Theories*, 1977, University of Illinois Press, p. 63.

21 Machamer, Peter, „A Brief Historical Introduction to the Philosophy of Science„ in *The Blackwell Guide to the Philosophy of Science*, 2002, Blackwell Publishers, p. 3

22 Kutasov, David, Mariño, Marcos, and Moore, Gregory, "Some exact results on tachyon condensation in string field theory", Journal of High Energy Physics, October, 2000. The field is entirely mathematical, as the cited work is presented as an example.

23 Suppe, Frederick, *The Structure of Scientific Theories*, 1977, University of Illinois Press, p. 289.

24 This is not theism or theology specifically, but metaphysics generally. The concept of "religion" itself is a contrivance of the rationalist era for the purpose of isolating Christianity from science. See William Cavanaugh, *The Myth of Religious Violence*, chapter 2, "The Invention of Religion."

25 Newton-Smith, W..H., editor, *A Companion to the Philosophy of Science*, 2001, Blackwell Publishers, p. 288.

26 Ibid, p. 289

27 Clouser, Roy, *The Myth of Religious Neutrality*, Second edition, 2005, Notre Dame Press. Religious" should not be confused with "religion" or "faith". But even though they are not interchangeable there is a clear conflict between the two because they possess different first principles. Theism has a personal deity and naturalism has impersonal principles. It is those impersonal principles, things which are taken as "unconditionally, non-dependently real," which places naturalism in its position of deity (p. 19).

28 Gould, Stephen Jay, *The Structure of Evolutionary Theory*, 2002, Belknap Press (Harvard), p. 725.

29 Machamer, Peter, and Silberstein, Michael, *The Blackwell Guide to the Philosophy of Science*, 2002, Blackwell Publishers, p. 55, Carl F. Craver's essay „Structures of Scientific Theories„

30 Ibid, p. 65

31 Don't let the term confuse you. The model for doing science is not just empirical model but now other types of models.

32 Ibid, p. 67

33 Abstract (object-oriented) languages such as Java and C++ are designed around classes. These classes are composites of both raw information and code. They may be cascaded so that Class A gains (inherits or implements) its data and code structures from parent classes. The abstractions in an application can become quite involved but would be otherwise too difficult or time-consuming to either develop or maintain without the availability of class structures.

34 Geier, David A., King, Paul G., and Geier, Mark R., "Influenza Vaccine: Review of Effectiveness of the U.S. Immunization Program, and Policy Considerations,,, *Journal of American Physicians and Surgeons*, Volume 11, Number 3, Fall 2006. This study addresses concerns of both individual effectiveness and population-wide effectiveness.

35 Suppe, Frederick, *The Structure of Scientific Theories*, 1977, University of Illinois Press, pp 64-152, quoting Kuhn: "Moreover, when one does look at how a scientist proposes or discovers these laws, theories, and hypotheses, one finds that he is not looking for anything like the physically interpreted deductive system of the Received View wherein his data are derivable consequences. Rather his initial search is for an explanation of the data – for a "conceptual pattern in terms of which his data will fit intelligibly along better-known data."

36 Ibid, p. 69

37 "... Kant (as we have seen) clearly states, in § 29 of the Prolegomena (the very passage where he gives his official "answer to Hume"), that there is a fundamental difference between a mere "empirical rule" (heat always follows illumination by the sun) and a genuine objective law (the sun is through its light the cause of heat) arrived at by adding the a priori concept of cause to the merely inductive rule. Any law thus obtained is "necessary and universally valid," or, as Kant also puts it, we are now in possession of "completely and thus necessarily valid rules."
http://plato.stanford.edu/entries/kant-hume-causality/

38 "Every genuine test of a theory is an attempt to falsify it, or to refute it. Testability is falsifiability; but there are degrees of testability: some theories are more testable, more exposed to refutation, than others; they take, as it were, greater risks.,,
http://www.stephenjaygould.org/ctrl/popper_falsification.html

39 Cleland, Carol E., "Historical science, experimental science, and the scientific method," *Geology*, November 2001, p. 987.

40 ibid, p. 988

41 I found that those of my friends who were admirers of Marx, Freud, and Adler, were impressed by a number of points common to these theories, and especially by their apparent explanatory power. These theories appear to be able to explain practically everything that happened within the fields to which they referred. The study of any of them seemed to have the effect of an intellectual conversion or revelation, open your eyes to a new truth hidden from those not yet initiated."
http://www.stephenjaygould.org/ctrl/popper_falsification.html

42 https://plato.stanford.edu/entries/scientific-objectivity/#FeyTyrRatMet, July 11, 2017
"The drawbacks of an objective, value-free and method-bound view on science and scientific method are not only epistemic. Such a view narrows down our perspective and makes us less free, open-minded, creative, and ultimately, less human in our thinking (Feyerabend 1975: 154). It is therefore neither possible nor desirable to have an objective, value-free science (cf. Feyerabend 1978: 78–79). As a consequence, Feyerabend sees traditional forms of inquiry about our world (e.g., Chinese medicine) on a par with their Western competitors. He denounces appeals to objective standards as barely disguised statements of preference for one's own worldview:

There is hardly any difference between the members of a "primitive" tribe who defend their laws because they are the laws of the gods [...] and a rationalist who appeals to "objective" standards, except that the former know what they are doing while the latter does not. (1978: 82)"

43 https://defenseofaith.wordpress.com/2015/12/03/presuppositional-objectivity/, July 11,2017

"For Van Til, objectivity in the Christian worldview is not a matter of having no presuppositions (and letting a pretended neutral reason to find the pretended external truth, which is actually organized by the subjective mind of man), but a matter of having the right presuppositions – that is, having the divine point of view gained through revelation."

Quoted from Bahnsen, Greg L., *Van Til's Apologetic*, P & R Publishing, 283-86

44 See Noll, Mark, *The Civil War as a Theological Crisis* for a fuller treatment of the global uniqueness of the theological South.

45 In *Black Mass, Apocalyptic Religion and the Death of Utopia*, John Gray presents the historic relationship between modern liberalism and certain Christian theological structures.

46 The Marxist theory of history, in spite of the serious efforts of some of its founders and followers, ultimately adopted this soothsaying practice. In some of its earlier formulations (for example in Marx's analysis of the character of the "coming social revolution") their predictions were testable, and in fact falsified. Yet instead of accepting the refutations the followers of Marx re-interpreted both the theory and the evidence in order to make them agree. In this way they rescued the theory from refutation; but they did so at the price of adopting a device which made it irrefutable.

McGrew, Timothy McGrew, Alspector-Kelly, Marc, Allhoff, Fritz, editors, *Philosophy of Science: An Historical Anthology*, Wiley-Blackwell, 2009, p. 474

47 That is, science as a product of the skepticism of the Rationalist movement and at the same time the eclectic combination of industrialism's mechanical view of life and postmillennialism's optimistic outlook for a better future.

48 (Recall the definitions of science: It is an explanatory model. It simply is not empirical science, at least not as a whole.)

49 Darwin, Charles, *The Origin of Species*, http://www.talkorigins.org/faqs/origin/introduction.html

50 Mayr, Ernst, *What Evolution Is*, Basic Books, 2002, p. 11

51 ibid, p. 76

52 ibid, p. 229

53 ibid

54 Ibid, p. 263

55 Gould, Stephen Jay, *The Structure of Evolutionary Theory*, 2002, Belknap Press, p. 116

56 Ibid, p. 121

57 ibid, p. 8. Gould was a seeking a more consistent application of Mayr's theory.

58 Prothero, Donald R., *Evolution: What the Fossils Say and Why It Matters*, p. 147

59 Ibid, p. 154ff

60 Ibid, p. 134ff

61 ibid

62 Ibid

63 Ibid, p. 4

64 http://www.merriam-webster.com/dictionary/hypothetico%E2%80%93deductive, cited 8/6/2016

65 https://explorable.com/hypothetico-deductive-method, cited 8/6/2016

66 Dawkins, Richard, *The Selfish Gene, 30th Anniversary Edition*, 2006, Oxford University Press, p. 2
"They made the erroneous assumption that the important thing in evolution is the good of the species (or the group) rather than the good of the individual (or the gene)." Dawkins would argue that the community always begins with the benefit of one individual and thus it all traces back to that one.

67 Ibid, p. ix

68 One text is *iGenetics: A Mendelian Approach* by Peter J Russell or for hands-on education there is the *Innovating Science Mendelian Genetics Kit* from Innovating Science.

69 Huxley, Julian; Wells, H. G.; Wells, G. P., *The Science of Life*, 1931, Doubleday, Doran & Company, Inc., Vol. II, p. 459ff.

70 Ibid, p. 486

71 Ibid, p. 490

72 Ibid, p. 507

73 http://www.fhcrc.org/en/news/releases/2013/05/harmit-malik-howard-hughes-medical-institute-investigator.html cited July 20, 2013.

74 Mayr, Ernst, *Animal Species and Evolution*, Cambridge: Belknap Press of Harvard University Press, 1963, p. 586

75 a b Valverde P, Healy E, Jackson I, Rees JL, Thody AJ (1995). "Variants of the melanocyte-stimulating hormone receptor gene are associated with red hair and fair skin in humans". *Nature Genetic*l 11 (3): 328–30. doi:10.1038/ng1195-328. PMID 7581459.

76 Dawkins appears to think of all behavior as inherently selfish. Even altruism he links to self-interest – bear a little pain now for greater ease later, as with herds and flocks of birds. See chapter 10 of The Selfish Gene.

77 Gould, Stephen Jay, *The Structure of Evolutionary Theory*, 2002, Belknap Press, p. 116

78 Ibid, p. 755

79 Ibid, p. 761

80 Shapiro, James, *Evolution, A View from the 21st Century*, 2011, New Jersey: FT Press Science, p. 1-2

81 Ibid, p. 137

82 Ibid, p. 128

83 Ibid, p. 134

84 Ibid, p. 138 "Over time it came to be unchallenged conventional wisdom that cognitive, goal-oriented processes have to be relegated to the realms of unscientific fancy and religion."

85 Ibid, p. 7

86 Prothero, op cit, p. 133-134

87 Ibid

88 Nelson, Paul A., *Biology and Philosophy 11*: 493-517, 1996, Netherlands: Kluwer Academic Publishers.

89 Barzun, op cit, p. 15

90 Barzun, op cit, p. 107ff

91 See *What Darwin Got Wrong* by Jerry Fodor and Massimo Piattelli-Palmarini and *Mind and Cosmos* by Thomas Nagel.

92 Barzun, op cit, p. 109

93 Huxley, Julian; Wells, H. G.; Wells, G. P., *The Science of Life*, 1931, Doubleday, Doran & Company, Inc., Vol. II, p. 432ff.

94 Coyne, Jerry A., *Why Evolution is True*, Viking, 2009, p. 58-59

95 Grant, Peter, and Grant, Rosemary, "Adaptive radiation of Darwin's finches: Recent data help explain how this famous group of Galapagos birds evolved, although gaps in our understanding remain.", in American Scientist, March-April 2002 v90 i2 p130(10) Also available at http://chiron.valdosta.edu/jbpascar/Courses/Biol1010/ExtraCreditActivities/American%20Scientist%20Online%20-%20Adaptive%20Radiation%20of%20Darwin's%20Finches.htm

96 http://news.yahoo.com/game-changer-evolution-african-bones-140125430.html Cited December 6, 2013

97 Coyne, op cit, p. 236

98 Fodor, Jerry, and Piatelli-Palmarini, Massimo, *What Darwin Got Wrong*, Farrar, Straus and Giroux, 2010, p. xiv

99 Ibid, p. 3

100 Williams, Michael, *Problems of Knowledge: A Critical Introduction*, 2001, Oxford University Press, p. 226

101 Coyne, Jerry A., *Why Evolution is True*, Viking, 2009, p. 92

102 Prothero, Donald R., *Evolution: What the Fossils Say and Why It Matters*, p. 15

103 Plantinga, Alvin, *Warrant and Proper Function*, 1993, Oxford University Press, p. 46

104 The terms univocal and evidential are being employed interchangeably here as are analogical and presuppositional. Though that is not as precise as it ought to be, there is a point in their relationship that is important to follow. All evidential arguments are univocal because they come with the assumption that the evidence speaks for itself.

105 ibid, pp. 1111-1112

106 ibid, p. 1081 ff

107 ibid, p. 1112

108 Ibid, p. 1125-1126

109 Ibid, p. 1126

110 Gould, Stephen Jay, *The Structure of Evolutionary Theory*, Belknap, 2002, p. 1110ff

111 Coyne, Jerry A., *Why Evolution is True*, Viking, 2009, p. 138

112 ibid, p. 139

113 Theologians and historians sometimes make this same mistake. Arguing from their systematic, the plain reading of historical documents is often re-interpreted [sometimes correctly, sometimes incorrectly] as it is driven by the systematic that is dominant in the mind of the interpreter. This is why we have denominations. This also happened recently in Egyptian historical studies and the dating of the

various kingdoms. In short it is not a problem unique to the scientific community. Colin Renfrew and Peter James have leveled some strong criticism of the "Sothic cycle" method of Egyptian dating.

114 Coyne, Jerry A., *Why Evolution is True*, Viking, 2009, p. 9

115 Ibid, p. 15

116 Ibid, p. 18-19

117 Coyne, Jerry, *The Nation*, "The Improbability Pump" , May 10, 2010, also available at http://www.thenation.com/article/improbability-pump

118 Coyne, op cit, p. 244-245

119 Barzun, op cit, p. 103-104

120 Coyne, Jerry A., *Why Evolution is True*, Viking, 2009, pp. 77-78

121 Isaak, Mark, *The Counter-Creationism Handbook*, University of California Press, 2007, pp. 83-84

122 Gould, Stephen Jay, *The Structure of Evolutionary Theory*, Belknap, 2002, p. 367

123 Ibid, p. 353

124 Mayr, Ernst, *What Evolution Is*, 2001, Basic Books, pp. 29-30

125 Gould, Stephen Jay, *The Structure of Evolutionary Theory*, 2002, Belknap Press (Harvard), p. 1025-1026

126 Ibid, 1027

127 Gould, Stephen Jay, "Spin Doctoring Darwin," *Natural History*, July 1995, quoted by Glick, Thomas, *What about Darwin?: All Species of Opinion from Scientists, Sages, Friends, and Enemies Who Met, Read, and Discussed the Naturalist Who Changed the World*, May 25, 2010, Johns Hopkins University Press, p. 154

128 Dawkins, Richard, The Selfish Gene, 30th Anniversary Edition, 2006, Oxford University Press, p. 36

129 Yockey, Hubert P., *Information Theory, Evolution, and the Origin of Life*, 2005, Cambridge, p. 23-24.

130 Oliphint, K. Scott, *Covenantal Apologetics*, Crossway, 2013, p. 105-6.

131 Fodor and Piattelli-Palmarini (op cit 167-168) deal with the problem of scientific epistemology. The Darwinian basis for knowledge. Quoting Campbell, the problem revolves around the very beginning of and accounting for knowledge. If it is all evolved, as Campbell states, then the argument is circular and accounts for itself. Thus the ideas that we might think are merely materialistic cognitive processes. They find this explanation inadequate. What remains is that shared knowledge and consensus end not with transcendent valuations but with individual brains, not minds. Shared belief ends in coincidence.

132 In this example the naturalist is not appealing to evidence but is also employing a presuppositional argument. All presuppositional arguments are by their nature circular. The issue is that the naturalist will make a claim about appealing to evidence when in fact presuppositional arguments are quite common.

133 Oliphint, K. Scott, *Covenantal Apologetics,* 2013, Wheaton: Crossway.

134 Ibid, p. 66-67

135 Chapter 7 of Alvin Plantinga's *God and Other Minds* provides a thorough examination of the logical issues faced by the verificationist. He confronts the earlier position of Anthony Flew that statements about God cannot be valid because, denying nothing, they can assert nothing. Plantinga works through the issue by proposing, among other things, that statements about God do deny opposites as God is exclusive from other propositions such as naturalism.

136 This is true at least in general. The problem of other minds, and also other beings, is one that persists, though by consensus we accept that other minds exist. For the sake of convention it seems better to accept that others exist. See A. Plantinga *God and Other Minds*.

137 Coyne, Jerry, "You can't prove a negative," from his personal *blog* „Why Evolution is True", cited October 14, 2013, http://whyscienceistrue.wordpress.com/2013/10/14/you-cant-prove-a-negative/

138 Mayr, op cit, p. 76

139 Ibid, p. 110

140 Ibid, p. 114

141 www.oocities.org/ginkgo100/faq.html

142 Gould, op cit, 742

143 Coyne, Jerry, *The Nation*, "The Improbability Pump", May 10, 2010, also available at
http://www.thenation.com/article/improbability-pump#axzz2aFZhvfsh

144 ibid

145 Nelson, Paul, "The Role of Theology in Current Evolutionary Reasoning" in *Biology and Philosophy*, 11: 493-517, 1996, Netherlands: Kluwer Academic Publishers.

146 Nelson, Paul, "Jettison the Arguments, or the Rule? The Place of Darwinian Theological Themata in Evolutionary Reasoning, from "Naturalism, Theism and the Scientific Enterprise: An Interdisciplinary Conference at the University of Texas," February 20, 1997, from http://www.discovery.org/a/104 cited on Sept. 27, 2017

147 Dilley, Stephen, "Charles Darwin's use of theology in the *Origin of Species*" in *British Society for the History of Science* 2011, doi:10.1017/S000708741100032X

148 We can argue whether mutation and adaptation amounts to evolution or whether this research might have been accomplished without a naturalistic framework. The point here is that the framework of naturalistic evolution was simply the one at work in this research.

149 Barzun, op cit, p. 15-16.

150 See *Darwin Day In America: How Our Politics and Culture Have Been Dehumanized in the Name of Science* by John G. West, *From Darwin to Hitler: Evolutionary Ethics, Eugenics, and Racism in Germany* by Richard Weikart, *Margaret Sanger's Eugenic Legacy: The Control of Female Fertility* by Angela Franks, and http://eugenicsarchive.org for an exploration into the history and scope of the eugenics movement.

151 http://www.nyu.edu/projects/sanger/webedition/app/documents/show.php?sangerDoc=101807.xml

152 Sanger, Margaret, *Woman, Morality, and Birth Control*, New York: New York Publishing Company, 1922. Page 12.

153 Sanger, Margaret, *Woman and the New Race*, 1920, republished by CreateSpace Independent Publishing Platform (May 20, 2013). Page 34.

154 Gershman, Bennett, "Three Generations of Imbeciles are Enough," http://www.huffingtonpost.com/bennett-l-gershman/eugenics-sterilization-anti-choice_b_1227929.html, cited October 1, 2013.

155 It is not a reach to suggest that the naturalistic scientist is here again appealing to Christian morality. There is a sense of guilt which accompanies this issue. The response to this guilt is to here argue that there is nothing to be guilty about.

156 https://stanford.library.sydney.edu.au/archives/spr2007/entries/sociobiology/ Cited 2/28/2018

157 Barzun, Jacques, *Darwin, Marx, Wagner: Critique of a Heritage, Second Edition with a new Preface,* Phoenix: University of Chicago Press, 1981, p. 15

158 William Cavanaugh develops this trend in *The Myth of Religious Violence* where he documents the change in language of the use of the term "religion" and how it became separated from Reason and thus from progress. Progress in the 19th century was being dissociated from the church and tied more tightly than ever to the increasing power of the nation to effect progress. This is the "progressive" movement.

159 Dilley, Stephen, *British Society for the History of Science,* 2011, doi:10.1017/S000708741100032X, p. 3. Dilley quotes Darwin's use of theology as supporting evidence. Whether or not Darwin accepted the truth value of his statements or whether they were intended merely to persuade the religious reader remains open. In either case, though, theological constructs remain a part of the history of his theory structure.

160 Gould, Stephen Jay, "Evolution as Fact and Theory," May 1981: from *Hen's Teeth and Horses Toes,* New York: W. W. Norton & Company, 1994, pp. 253-262.

161 http://en.wikipedia.org/wiki/Saltationism

162 Gould, Stephen Jay, *The Structure of Evolutionary Theory,* 2002, Belknap Press (Harvard), p. 1008-1009. Gould defends his punctuated equilibrium against the charge of saltationism brought by Daniel Dennett. Though Dennett employs the fuller claim of "saltationism" he is right to note that Gould is making room for changes that must be so rapid that they become immediate visible. Hence my prefix of "soft" appears to be a more accurate criticism.

163 "Marxism, for a Popperian, is scientific if the Marxists are prepared to specify facts which, if observed, make them give up Marxism. If they refuse to do so, Marxism becomes a pseudoscience. It is always interesting to ask a Marxist, what conceivable event would make him abandon his Marxism. If he is committed to Marxism, he is bound to find it immoral to specify a state of affairs which can falsify it. Thus a proposition may petrify into pseudoscientific dogma or become genuine knowledge, depending on whether we are prepared to state observable conditions which would refute it." Lakatos, Imre, "Science and Pseudoscience," from *Philosophy of Science: An Historical Anthology,* McGrew, Timothy, Alspector-Kelly, Marc, and Allhoff, Fritz, Editors, 2009, Wiley-Blackwell, p. 561.

www.ingramcontent.com/pod-product-compliance
Lightning Source LLC
LaVergne TN
LVHW051648080426
835511LV00016B/2566